Introduction To Map Projections

Porter W. McDonnell, Jr.

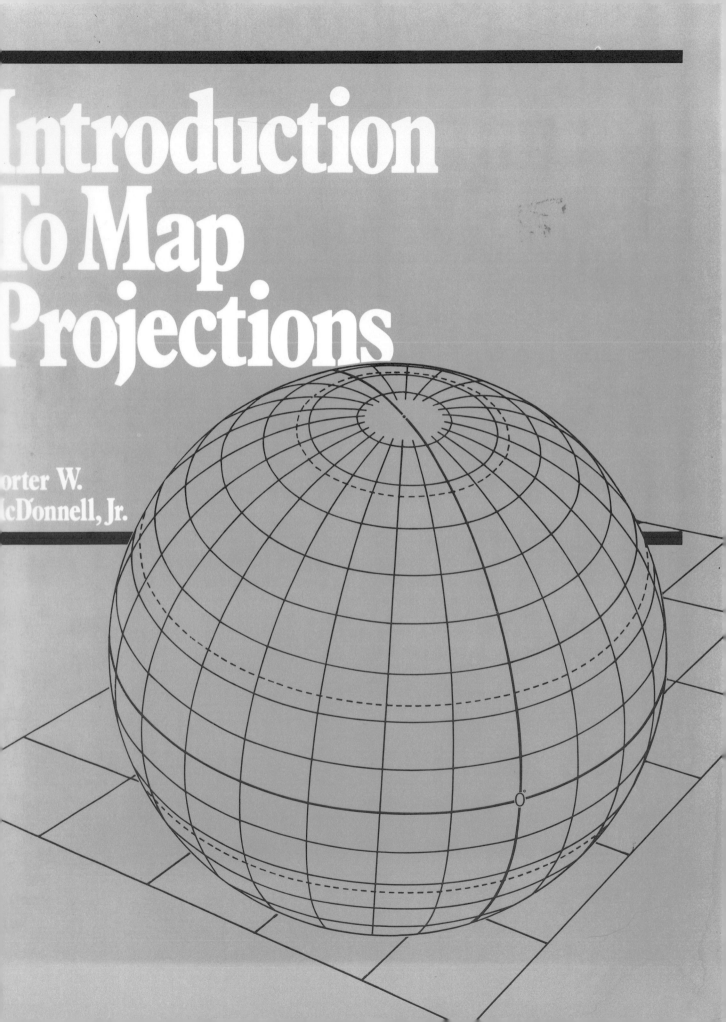

INTRODUCTION
TO MAP PROJECTIONS

INTRODUCTION
TO MAP PROJECTIONS.

Porter W. McDonnell, Jr.

Associate Professor of Engineering
The Pennsylvania State University
Mont Alto, Pennsylvania

MARCEL DEKKER, INC.   NEW YORK AND BASEL

Library of Congress Cataloging in Publication Data

McDonnell, Porter W   [Date]
    Introduction to map projections.

    Includes index.
    1. Map-projection.   I. Title.
GA110.M25   526.8   79-13560
ISBN 0-8247-6830-2

MARCEL DEKKER, INC.
270 Madison Avenue, New York, New York 10016

Current printing (last digit):
10 9 8 7 6 5 4 3 2 1

PRINTED IN THE UNITED STATES OF AMERICA

This book is intended to serve as a text or reference on map projections for students of geography, surveying, and cartography. Other aspects of cartography such as map drafting, lettering, symbolization, map reproduction, and the design of thematic maps are not covered here. Five good references on these related matters are listed at the end of Chap. 1.

The chapter headings show that the various projections are grouped differently than in other books on the subject. Instead of listing them according to the projection surfaces commonly used in describing most projections (cylinder, cone, and plane), this text uses certain shared characteristics as the chapter titles. This arrangement is intended to emphasize similarities rather than differences, and constitute an aid to learning.

Each chapter includes a set of problems (for a total of 82). Only a few involve drafting, but many are suitable for conversion to map-drafting exercises. Some instructors will want to use them in that way. Chapter 11, which introduces map compilation, may be assigned out of sequence in such cases.

The coverage is intentionally less mathematical than that of Map Projections for Geodesists, Cartographers and Geographers by P. Richardus and R. K. Adler (North-Holland Pub., Amsterdam, 1972), which is intended for students at the masters degree level. Here the mathematics only rarely involves calculus or spherical trigonometry and neither is essential to the understanding of the book. The datum used, in most cases, is the sphere rather than a spheroid.

All numerical examples and problems are in metric units, but occasional references to miles, feet, and inches are included for those preferring to teach with English units. Where the problems list alternate units the answers or partial answers given refer only to the SI units.

Chapter 10 includes a more complete description of the important Universal Transverse Mercator and Universal Polar Stereographic grid systems than is available in any other cartography textbook to come to the author's attention. The Appendixes include a table listing the characteristics of the most important map projections, some supplementary material on Tissot's indicatrix, and several programs for pocket calculators.

The book was written originally for associate degree students in surveying technology but includes enough material to be useful as a concise text or lab manual in any other college programs where cartography must be introduced. As usual, it was the lack of a suitable book that inspired the writing of this one.

No preface is complete without an acknowledgment by the author of the obvious fact that he could not have completed the project without great quantities of help from others. Joseph F. Douglas, George N. Payette, and Vernon L. Shockley of Penn State were the administrators who recognized that the gleam in the author's eye was genuine. They gave him some time in which to get the work started during the spring term of 1977. John Hychko and Gordon Bowker of Penn State, and William D. Brooks of Indiana State University were kind enough to use Xerox copies of the first draft in their classes. Very helpful review of parts of the first draft were provided by Judy M. Olson of Boston University, Olubodon O. Ayeni during a one-term appointment at Penn State, and John P. Snyder of CIBA-GEIGY Corp., whose spare-time activities have made him an authority on map projections while working in another field. Mr. Snyder also contributed the calculator program in Sec. A-4. Helpful in smaller ways were Bro. B. Austin Barry, Robert F. Maurer, and Kresho Frankich.

Illustrations were prepared by former student R. Craig Shuman, Jr., a graphic artist, except where they are specifically credited to others. All of the typing, from the first to the final draft, was done with precision and promptness by Noreen G. Verdier, and there was a generous amount of wifely support from Margaret McDonnell.

Porter W. McDonnell, Jr.

CONTENTS

CONTENTS   vii

THE GEOMETRY
OF MAP PROJECTIONS

1-1   INTRODUCTION

The earth is round; maps are flat. If a particular map is to show only a very small portion of the earth, such as a few city blocks, the roundness of the earth will be insignificant. If, on the other hand, a map is to show the Western Hemisphere, the roundness will present a major problem. Some kind of deformation will be necessary. (A large section of orange peel can only be flattened if it is stretched and torn.)

The map showing only a small area is often called a plan. The preparation of plans is the province of the land surveyor, city planner, photogrammetrist, engineer, or architect. He may describe his work as mapping, drafting, or topographic drawing. Making a map of the Western Hemisphere, in contrast, is the province of the geographer or cartographer, who may describe his work with any of the terms just mentioned but is more likely to call it cartography or cartographic drafting. Although cartography really includes map making at any scale, the term is commonly applied to the mapping of a large portion, or all, of the earth (or moon) where curvature of the surface is a factor. For such large areas (and small scales) the product is called a map or chart, the latter term being used if it is designed for navigational purposes. The preparation of a plan will involve a simple rectangular grid, whereas a map or chart commonly requires the selection of a suitable map projection to deal with the rounded nature of the earth.

Although the roundness of the earth will not be a factor in drawing a plan (a map of a limited area), the topic of map projections is nevertheless of importance to land surveyors. Increasingly, they are making use of plane coordinate systems which extend over hundreds of kilometers (or miles) even though the job at hand is small. Where a plan shows a limited area of land and is drawn as if the earth were flat, the data shown may be of such precision that a knowledge of map projections was needed in the survey computations.

This book will serve to introduce the subject of map projections to students of surveying, geography, and the specialty of cartography. Different kinds of students will use the background provided here in varying ways.

Other aspects of cartography (map design, reproduction, etc.) are left to other books [1, 2], as is the subject of survey drafting [3].

THE EARTH AND ITS GRATICULE

1-2 SIZE AND SHAPE OF THE EARTH

For most of the purposes of this book, and for small-scale mapping generally, the earth may be considered to be a sphere with a radius of 6,370 km (or about 3,960 miles). Actually, the dimension is greater across the Equator than it is from top to bottom (pole to pole). (The radius varies from 6,378 km at the Equator to 6,356 km along the polar axis.) The earth's surface area is equal to that of a sphere having a radius of about 6,370 km. For use in computing map projections, it is convenient to memorize this radius as 637,000,000 cm.

The meter originally was intended to be 1 ten-millionth of the distance from the Equator to the pole. Using the sphere of equal area just described, this distance is one-quarter of $2\pi R$, or 10,006 km.

For large-scale mapping and in geodetic surveying, the true shape of the earth may have to be considered. Some examples of this kind are discussed in Chaps. 3, 6, and 9, where a spheroid is used instead of a sphere. Figure 1-1 is a flow chart in which the choice of a datum (sphere or spheroid) is shown as the first step in the evolution of a map projection.

1-3 MAP PROJECTION

If all points within some large portion of the earth (the Western Hemisphere, for example) are to be represented on a flat map, two transformations of the sphere or spheroid must take place. First, there must be a scale reduction to make such a huge area fit into the limits of the paper; second, there must be some systematic way of deforming the rounded surface of the sphere or spheroid to make it flat. (See Fig. 1-1.)

It is very useful to think of these operations as always being done in two steps, in the order mentioned. First, the full-sized sphere is greatly reduced to an exact model called a globe. Second, a map projection is generated in some way to convert all or part of the globe into a flat map. There are an infinite number of ways, literally, of accomplishing this second step.

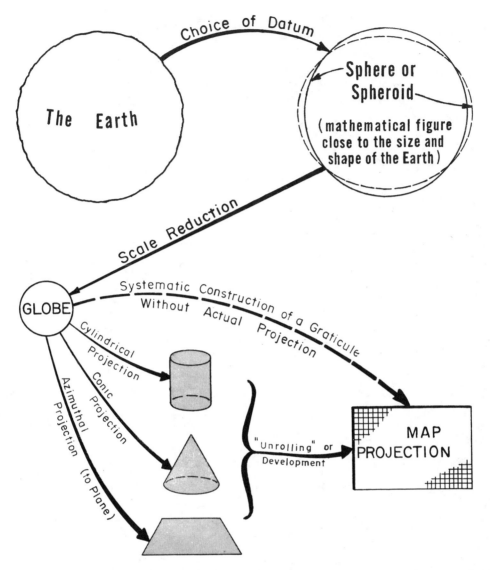

FIG 1-1   Evolution of a map projection.   In some cases a geometric projection
to a developable surface is involved, but usually the term cylindrical, conic, or
azimuthal is used to classify a projection which only resembles such a case.   The
dashed arrow shows this possibility.

Certain reference lines and points have been established on the earth.   The
Equator and the two poles are known to all.   See Fig. 1-2.   The lines running
north and south, from pole to pole, are meridians.   One of them, passing through
Greenwich, England, has been chosen arbitrarily to be the Prime Meridian.   It
is assigned an angular value of 0°.   The other meridians are identified by their
angular distance east or west of the Prime Meridian.   The meridian through
Pittsburgh, for example, is 80° west of Greenwich and is said to have a longitude

FIG. 1-2   The network of meridians and parallels (called the graticule). Pittsburgh is located at ϕ = 40°N and λ = 80°W. (Copyright by Denoyer-Geppert Co.,
Chicago; used by permission.)

of 80°W. (The line itself is the meridian, its spherical coordinate is the longitude.) The Greek letter lambda (λ) is used for longitude. The lines crossing all
meridians at right angles and running parallel to the Equator are called <u>parallels</u>.
They are identified by their angular distance north or south of the Equator. The
parallel passing through Madrid, Pittsburgh, and Peking is at 40°N. That value

is its <u>latitude</u>. The line is often called the 40th parallel if there is no chance of confusing it with the parallel at 40°S. The Greek letter phi (ϕ) is used for latitude.

The network of meridians and parallels is called the <u>graticule</u>. When the sphere or spheroid is reduced in size, as shown in Fig. 1-1, the graticule becomes, of course, a reference network for all points on the globe just as it is on the earth itself.

The map to be produced will be two-dimensional. Points on the map sheet will have x and y positions based on some rectangular system of reference. The process of map projection is the systematic transformation of all spherical coordinates (ϕ and λ) of the globe into corresponding rectangular coordinates (x and y) of the map. Mathematically,

$$x = f_1 (\phi, \lambda)$$

$$y = f_2 (\phi, \lambda)$$

meaning that x and y positions on the map are functions of ϕ and λ. The functions must be unique (so that a particular point will appear at only one position on the map), they must be finite (so that a particular point will not appear at infinity and be unplottable), and they must be continuous (so that stretching or shrinking of features may occur but there will be no gaps). Projections do exist in which the functions are not finite for the entire globe.

In a few cases, the x and y positions may be obtained by imagining an intermediate step involving a cylinder, cone, or plane as shown in Figs. 1-1 and 1-3. To illustrate this type of projection, imagine a ray of light projecting radially from the center of the globe to one of the tangent surfaces (Fig. 1-3). A point on the globe having a certain ϕ and λ may be transferred to the surrounding surface, which then may be "unrolled" or <u>developed</u> to form a plane map. More commonly the functions used to get x and y positions are purely mathematical and do not involve a developable surface. Many of these mathematical concoctions bear some resemblance to the geometrically projected cases shown in Fig. 1-3. The cylinder, cone, and plane thus form a common basis for classifying a large number of projections. A projected graticule is classified as <u>cylindrical</u> if it takes on a rectangular appearance, <u>conic</u> if it looks fan-shaped, and <u>azimuthal</u> if it resembles a map projected directly to a plane. The term azimuthal refers to the property that azimuths (or bearings or directions) from the central point to other points are not deformed during the process of projection. This term will be discussed further in Sec. 1-10 and in Chap. 2. An example of the cylindrical group is the well-known Mercator projection (Chap. 6).

If a projected graticule has only a slight similarity to the geometrically projected cases it may be classified as pseudocylindrical, pseudoconic, or pseudo-azimuthal. Figure 1-15 shows an example of a pseudocylindrical projection. It is not rectangular in appearance but the parallels are all horizontal, suggesting a relationship to the cylindrical group.

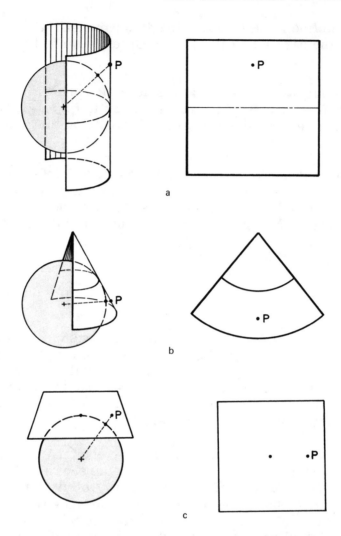

FIG 1-3 Developable projection surfaces. a. Cylinder tangent to a globe at the
Equator, and the developed or "unrolled" map. b. Cone tangent to a globe along
a parallel, and the developed or "unrolled" map. c. Plane tangent to a globe at
the north pole, and part of the resulting map.

MATHEMATICAL PRINCIPLES

1-4 SCALE

When a large area, such as the Western Hemisphere, is shown on a small
sheet of paper, the result is said to be a small-scale map. A large-scale map,
of course, is the opposite; an example would be the map of a few city blocks on
a large sheet. The large-scale map is generated from a relatively large globe.

The usual way of expressing scale in numerical terms is by a dimensionless ratio or representative fraction (RF).

$$\text{RF scale} = \frac{\text{globe distance}}{\text{earth distance}} \qquad (1)$$

If 200 km (124.3 miles) is represented on a globe as 1 cm, the RF scale is

$$\frac{1.00 \text{ cm}}{200 \text{ km}} = \frac{0.0100 \text{ m}}{200,000 \text{ m}} = \frac{1}{20,000,000}$$

Distances on the globe are 20,000,000 times smaller than they are on the earth itself. (Two-dimensional surfaces, or areas, are reduced in both dimensions and thus are smaller by a factor of $20,000,000^2$ or 4.00×10^{14}, but it is the linear scale that is usually stated.) This scale, often called the principal scale, may be written as 1:20,000,000, for convenience. If the radius of the generating globe R is known, the RF scale is equal to R/6,370 km converted to a dimensionless ratio with a numerator of unity.

Scale also may be expressed in unit equivalents. In the case mentioned, this would be 1 cm represents 200 km or

1 cm = 200 km

in which it is understood that the smaller unit is a globe distance and the larger unit is an earth distance. It is standard practice to assign unity to the smaller unit rather than to say 1 km = 0.005 cm, or 1 km = 0.05 mm. Other examples of unit equivalent scales are 1 in. = 300 miles, and 1 in. = 2,000 ft.

Scale may be shown graphically, with convenient multiples of earth distances marked off along a bar. In the first example above, scale divisions equivalent to 100 or 500 km might be used. The size of a 500-km division would be 500,000 m/ 20,000,000 = 0.025 m or 2.5 cm. Graphic scales, being pictorial, are very helpful to the map user.

1-5 SCALE FACTOR

All dimensions of the earth are reduced proportionately when the earth is imagined to be reduced to a globe. Some dimensions do not undergo any further change as the surface of the generating globe is projected to become a map. Figure 1-3a shows the case of a cylinder tangent to the globe at the Equator. As the cylinder is unrolled, or developed, the Equator maintains its original length. Such a line is called a standard line, or a line of exact scale. It is said to have a scale factor of 1.000. If a certain line is doubled in length during the projection process, it is said to have a scale factor of 2.000. In equation form,

$$\text{Scale factor} = \frac{\text{map distance}}{\text{globe distance}} \qquad (2)$$

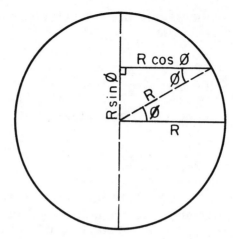

FIG 1-4   Cross section of a globe showing that the radius of any parallel of latitude is R cos ϕ and that the distance between its plane and that of the Equator is R sin ϕ.

The scale factor on any map will vary from point to point and may vary in different directions at the same point, being 1.000 along only standard lines or at a standard point. No map has a uniform scale. An RF, such as 1:20,000,000, applies to the generating globe itself and is correct for the map only where the scale factor is 1.000. The scale factor is also known by the name "particular scale."

1-6   THE MATHEMATICS OF THE SPHERE

Before studying the subject of map projections, it is important to understand the geometry of spheres.

If the radius of the globe is R, the circumference is $2\pi R$. That, of course, is the length of the Equator and of any meridian circle. The various parallels are shorter in circumference than the Equator. The north and south poles are really the 90th parallels in the Northern and Southern Hemispheres, but they have zero lengths.

The length of a particular parallel may be calculated by multiplying the length of the Equator by the cosine of the latitude. Using ϕ for latitude, the relationship is

Length of parallel = (length of Equator) (cos ϕ)   (3)

Figure 1-4 shows why the above statement is correct. It is a cross section of a sphere in which it can be seen that the radius of any parallel, in its own plane, varies with cos ϕ. The circumference of any parallel, in turn, is $2\pi R \cos \phi$. Obviously, the partial lengths of the parallels falling between any two meridians also varies with cos ϕ, becoming zero at the poles where cos ϕ is zero.

Example 1-1

On a globe with a radius of 30 cm, find the distance from Pittsburgh to Peking as measured along the 40th parallel.

Solution

The longitudes of the two cities are listed in some atlases. Also, it is possible to scale them from maps. For Pittsburgh, λ = 80°W and for Peking λ = 116°E. If one goes <u>east</u> from Pittsburgh the difference in longitude, called $\Delta\lambda$, will be 80° + 116° = 196°. Clearly, it will be shorter to go west, in which case the difference in longitude will be 360° - 196° = 164° and the distance will be

$$\frac{164°}{360°} \; 2\pi 30 \cos 40° = 65.8 \text{ cm}$$

In addition to seeing this solution as 164/360 of a circumference ($2\pi 30 \cos 40°$), it is helpful to view it in terms of radians. (In the radian system, an angle is equal to the ratio of the arc divided by the radius, and 2π radians corresponds to 360°.) The arc required in this problem is equal to a central angle in radians, 164° ($2\pi/360°$), times a radius, 30 cos 40°. The conversion of 164° to radians may be done more directly on some pocket calculators.

The surface area of a sphere is $4\pi R^2$. It happens that this area is exactly equal to that of a cylinder of the same diameter and height. The circumference of such a cylinder is $2\pi R$ and the height is 2R. The area is ($2\pi R$) (2R) = $4\pi R^2$, as stated for the sphere.

The surface area of the zone between any two parallels on a sphere may be found in a similar manner. It is equal to that of a strip on the surrounding cylinder having the same height (see Fig. 1-5). The height of the strip shown is R sin ϕ and the area, therefore, is

Area, Equator to parallel = ($2\pi R$) R sin ϕ = $2\pi R^2$ sin ϕ (4)*

Example 1-2

Assuming the earth to be a sphere with a radius of 6,370 km, find the area falling between the Tropic of Cancer (23.5°N) and the Arctic Circle (66.5°N), known as the Temperate Zone.

---

*Equation (4) (for the area of the zone between the Equator and any parallel) also may be found by integration. A narrow zone located at any latitude will have a width of R dϕ and a radius of R cos ϕ. Its area is ($2\pi R$ cos)(R dϕ) and the total area desired therefore is

$$\int_0^\phi 2\pi R^2 \cos \phi \, d\phi = 2\pi R^2 \sin \phi$$

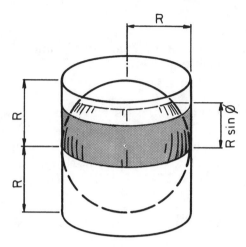

FIG 1-5   Globe surrounded by a cylinder of the same height 2R.   The surface area
of the shaded zone on the sphere is equal to that of the band on the cylinder having
the same height.

Solution

The plane containing the Arctic Circle will reach the surrounding cylinder at
a distance above the Equator equal to R sin ϕ = 6,370 sin 66.5° = 5,842 km.
The plane of the Tropic of Cancer will reach the cylinder at a distance of
6,370 sin 23.5° = 2,540 km.   The required area is $(2\pi 6,370)(5,842 - 2,540)$ =
132,200,000 km^2.   (Note that the answer is given to four significant figures
even though a calculator might give 9 or 10 digits.   Section 1-2 pointed out
that the radius of 6,370 km is only an average value having four significant
figures.   In fact a better average for the Temperate Zone might be different
in the fourth place.)

The term <u>great circle</u> is used to refer to any arc on the earth (or globe) formed
by a plane containing the center of the earth.   Each meridian is a great circle (or
actually half of one, running only from pole to pole).   The Equator is another example.   The shortest distance between any two points on the earth's surface is the
great circle route.   The shortest route between Pittsburgh and Peking is not the
one discussed in Example 1-1, but rather a great circle route running consider-
ably above the 40th parallel.

The shortest distance between two points may be found by first calculating the
central angle subtended by the two points (as measured in the plane of the great
circle) using an expression from spherical trigonometry.   In Fig. 1-6, if D is
the arc distance (central angle) between points A and B, ϕ_a is the latitude of A,
ϕ_b the latitude of B, and $\Delta\lambda$ is the difference of longitude between A and B, the
expression is

$$\cos D = (\sin \phi_a)(\sin \phi_b) + (\cos \phi_a)(\cos \phi_b)(\cos \Delta\lambda) \tag{5}$$

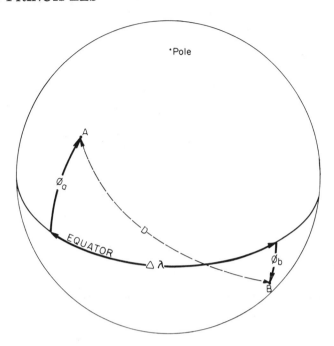

FIG 1-6   The terms in Eq. (5) include the desired arc distance D between points A and B, the latitudes ϕ_a and ϕ_b of the two points, and the difference in longitude $\Delta\lambda$.

Note that ϕ_a and ϕ_b must be expressed as plus or minus (north or south of the Equator), but that $\Delta\lambda$ may be the longitudinal difference in either direction (not necessarily less than 180°). The latter statement is correct because cos $\Delta\lambda$ is the same either way; for example, cos 20° = cos 340° and cos 200° = cos 160°. The arc distance D may be converted to a surface distance using the assumption that the earth is spherical. The length of each degree of a great circle is, of course, $2\pi R/360°$.

Example 1-3

Find the shortest distance from Pittsburgh to Peking on the globe of Example 1-1 (where R = 30 cm).

Solution

cos D = (sin 40°)(sin 40°) + (cos 40°)(cos 40°)(cos 164°)

      = -0.1509

D = 98.68°

Surface distance = $(2\pi 30/360°)(98.68°)$ = 51.7 cm

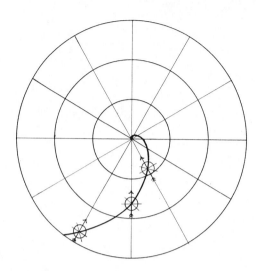

FIG 1-7   A loxodrome crossing meridians at a constant angle.   (From Ref. 4 with permission.)

The average length on earth of a degree on a great circle, using the expression $2\pi R/360°$, is 111. 2 km or 69. 1 statute miles.   For a 1-minute arc, the length is 1. 853 km or about a nautical mile.   (The international nautical mile is 1. 852 km. Slightly greater values also have been in use.)   The length of a 1-second arc is about 30. 9 m or 101 ft.

Although a great circle provides the shortest possible route between two points, it may be a difficult route to follow if navigating manually by compass.   In flying from Pittsburgh to Peking it would be simpler to go due west all the way than to follow a route with a constantly changing bearing.   (The great circle route would be northwest at first and southwest later.)   The route of constant bearing (or constant azimuth), in this case the 40th parallel, is called a <u>loxodrome</u> or <u>rhumb line</u>.   The parallels and meridians are loxodromes, of course, but in general a loxodrome is a route which crosses every meridian at the same angle (see Fig. 1-7).   The trip from Pittsburgh to Peking could follow a series of loxodromes which together would approximate the great circle route.   The pilot would change his bearing a number of times but not continually.   There is a map projection which has the valuable characteristic that straight lines drawn upon it are great circle routes (gnomonic, see Chap. 7).   There is another one on which straight lines are loxodromes (Mercator, see Chap. 6).

Another useful bit of geometry relates to the trapezoid (see Fig. 1-8).   Its area is

$$\text{Area} = \frac{b_1 + b_2}{2}h \tag{6}$$

A 1° quadrangle (bounded by parallels and meridians 1° apart) resembles a trapezoid but is on a curved surface instead of a plane.   The two bases are curved.   A

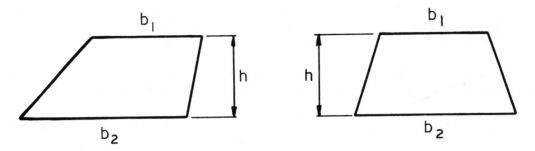

FIG 1-8 Examples of trapezoids (plane four-sided figures with two parallel sides). Both figures have the same area.

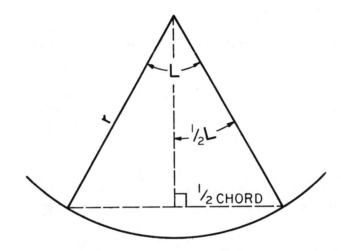

FIG 1-9 Relationships for computing chord lengths.

1-minute quadrangle is even more like a trapezoid because there is much less curvature involved. This idea is used in Chap. 5.

1-7 PLOTTING CIRCULAR ARCS

On some map projections a parallel of latitude will be drawn as part of a circle, as in Fig. 1-9. The longitudinal coverage $\Delta\lambda$ represented by the circular arc may be a full 360° or any lesser amount, such as 16° for a map of France.

The figure shows a central angle L which will be related to $\Delta\lambda$ but generally not equal to it. For example, a map may show all 360° of the 40th parallel as a circular arc in which L = 231°. The central angle L is related to $\Delta\lambda$ by a constant k as follows:

$$L = k \, \Delta\lambda \qquad\qquad\qquad (7)$$

In the case cited, L = (0.643) (360°) = 231°. For reasons which will become clear in the next few chapters, k is called the <u>constant of the cone</u>.

In this book the radius for drawing a circular arc will be called r (and R will be reserved for the radius of the globe as in earlier sections of this chapter).

In dealing with circles it will be useful to remember that a chord length may be calculated as follows, referring again to Fig. 1-9.

$$\sin 1/2 \ L = \frac{1/2 \ \text{chord}}{r}$$

$$\text{chord} = 2r \sin 1/2 \ L \tag{8}$$

This expression is useful in fields such as route surveying as well as in the study of map projections.

Often there is a need to calculate the x and y coordinates (or tangent offsets) of several points along a circular arc as shown in Fig. 1-10. If r = 50 cm the arc might be drawn using a beam compass, but if r = 100 cm it might be easier to plot a series of points and join them using a flexible curve.

As an example, assume that a particular parallel has a radius of 100.0 cm and that the meridians will cross it as intervals of 5.60 cm. (Fig. 1-10 is not drawn to scale.) The central angle for each 5.60-cm arc will be

$$L = \frac{5.60}{100.0} \ \text{radians} \quad \text{or} \quad \left(\frac{5.60}{100.0}\right)\left(\frac{360}{2\pi}\right) = 3.2086°$$

A plotting table may be prepared giving x and y coordinates from the point 0 where the central meridian meets the parallel. If five graticule points are needed, the table will be as follows:

Point no., n	x = 100.0 sin (3.2086°n), cm	y = 100.0 - 100.0 cos (3.2086°n), cm
0	0.00	0.00
1	5.60	0.16
2	11.18	0.63
3	16.72	1.41
4	22.21	2.50
5	27.64	3.89

These values may be plotted to both the left and right of the origin. If the meridians are spaced 4° apart, then, using Eq. (7), k must be 3.2086/4.00 = 0.8022.

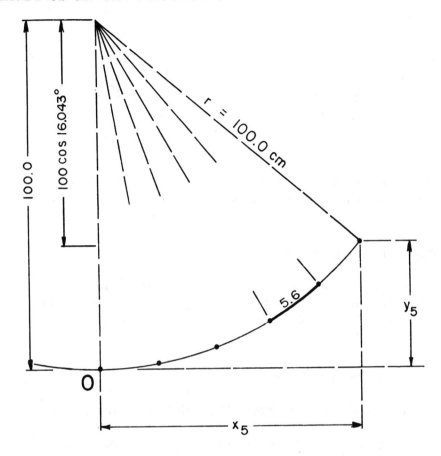

FIG 1-10   Tangent offsets for plotting point 5 along a circular arc.

CHARACTERISTICS OF MAP PROJECTIONS

It has been mentioned that an infinite number of map projections are possible. About 15 or 20 of them are in common use.   To aid the reader in understanding most of these projections, this book will group them according to properties they share.   (It is more common to classify them as conic, cylindrical, etc., as mentioned in Sec. 1-3.) The next five sections describe some of the important characteristics that some projections have in common.

1-8 STANDARD LINES

In the discussion of scale factor (Sec. 1-5), standard lines were defined as lines which do not change length when projected from a generating globe to a map. Some projections have only one standard line (such as the Equator in Fig. 1-3a), but others may have many. Several of the conic projections, for example, have two standard parallels.

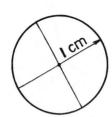 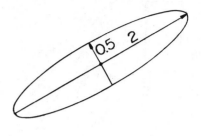

FIG 1-11 A circle on a globe, shown at left, will project as an ellipse on a map. An example from an equal-area projection is shown.

1-9 EQUIDISTANT PROJECTIONS

Although no map projection can offer a uniform scale, some of them have it in one direction. A projection may have a scale factor of 1.000 in the north-south direction, for example (all meridians are standard). This kind of projection will be described as equidistant.

1-10 AZIMUTHAL PROJECTIONS

Section 1-3 mentioned the characteristic which all azimuthal projections share, namely, that directions to all points with respect to a central point are not deformed during projection from globe to map. (The direction to a distant point is important in the operation of airports, seismographs, radio stations, etc.)

1-11 EQUAL-AREA PROJECTIONS

If the relative size of all features on a generating globe is maintained during the process of projection to a map, the projection is said to be equal-area (also equivalent or equiareal).

It has been pointed out (Sec. 1-5) that no map has a scale factor of 1.000 everywhere. If area is to be preserved but scale cannot be, then a given feature on a globe, such as a state, will have to be plotted with a scale factor greater than 1.000 in one direction and less than 1.000 in another. It can be shown that such "compensatory scale factors" on an equal-area projection will occur in perpendicular directions, often called the principal directions. If a circle with a radius of 1 cm is drawn upon the surface of a large generating globe, it will appear as an ellipse when projected to an equal-area map (see Fig. 1-11). If one semiaxis is reduced to 0.5 cm, the other will be increased to 2.0 cm, in order that the ellipse may contain the same area as the circle.

1-12 CONFORMAL PROJECTIONS

While equal-area projections have compensatory scale factors and allow a tiny circle, as just described, to be distorted in order to avoid a change in its area,

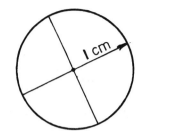

FIG 1-12 In conformal projections, a circle on a globe, at left, will project as another circle (a special kind of ellipse). A scale factor of 0.6 is illustrated.

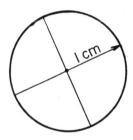
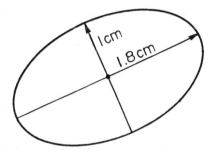

FIG 1-13 In a projection that is neither equal-area nor conformal, a circle on a globe, at left, will project as a noncircular ellipse of different size.

conformal projections have equal scale factors in all directions at any one point. A tiny circle is not distorted at all but is allowed to become simply a larger or smaller circle, depending upon its location on the map. Instead of preserving size, conformal projections preserve shape (see Fig. 1-12). They also are called orthomorphic projections.

If the shape of everything on a globe could be preserved on the map, as was that of the tiny circle in the preceding paragraph, the map would have a uniform scale, which is impossible. Thus it is correct to say that shapes of small features will be preserved in the course of projecting to the conformal map. Conformal projections are ideal for setting up plane coordinate systems for use in surveying because a surveyor's traverse is small in comparison to the portion of the earth on a particular system. Its angles will be the same when placed on the coordinate system as when they were measured in the field. There will be a single scale factor (generally not 1.000) that will apply to all distance measurements, unless the survey is of unusually great size or precision.

A map certainly cannot be both equal-area and conformal. It cannot have minimum and maximum scale factors at a point which are compensatory and yet equal. Some projections are neither equal-area nor conformal. The tiny circle referred to earlier is then distorted in both size and shape. See Fig. 1-13.

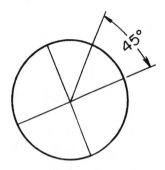 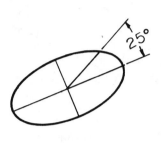

FIG 1-14 When a circle is projected as a noncircular ellipse, the angle between two radial lines will be deformed (except for the two perpendiculars which become the axes of the ellipse).

It may be shown that for any projection such a circle will invariably be transformed into an ellipse of some size. It will have a pair of axes (the principal directions) which remained perpendicular during projection. Unless the projection is conformal, other angular relationships at the point will be disturbed. Angular deformation at a point will be zero for the principal directions and will reach some maximum value for another pair of lines (see Fig. 1-14). Figure 1-15 shows how this maximum angular deformation varies over an equal-area projection of the world. Clearly the meridians and parallels do not always meet at right angles as they did on the globe.

It is evident in Fig. 1-15 that the pair of perpendiculars which remains perpendicular after projection is not necessarily the pair of graticule lines at a point. When it is that pair, as in cylindrical projections, the dimensions of the small ellipse are easy to compute. These calculations will be done in later chapters to illustrate the nature of the projection being studied.

The mathematics of how the tiny circle is deformed into an ellipse was developed by M. A. Tissot in 1881. Further discussion appears in Appendix C.

PROBLEMS

1-1 If a globe has a diameter of 25 cm, what is its RF scale? Also express the scale on a unit equivalent basis (1 cm = ?).

Partial Answer: 1 cm = 509.6 km

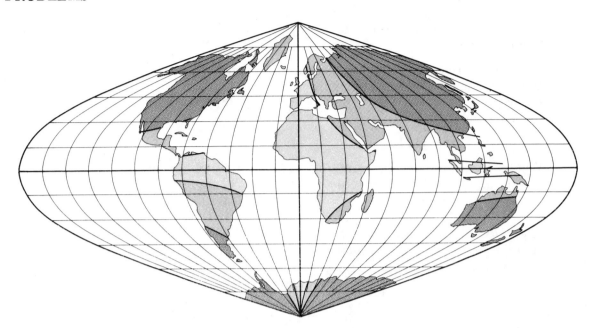

FIG 1-15 A world map showing lines of equal maximum angular deformation (10°
and 40°). Projection is sinusoidal, discussed in Chap. 5. (From Ref. 5 with permission.)

1-2 Find the height of Mt. Everest as it would appear on the globe in Prob. 1-1.
Express the answer in millimeters. (The elevation of the summit is 8848 m.)
Would a true-scale relief globe of this size be worthwhile?

  Partial Answer: Not worthwhile

1-3 Find the diameter of the 60th parallel on the globe of Prob. 1-1. How far is
this parallel from the Equator as measured along a meridian?

  Partial Answer: 13.09 cm to parallel

1-4 Which of the following map scales is largest?

  a. 1:5,000,000   b. 1/50   c. 1 cm = 10 km

1-5 On a large globe in a library, the distance between the 40th and 50th parallels
is found to be 5.32 cm. How wide should a 10° quadrangle be at the top and
bottom if it lies between these parallels? Consider it to be approximately a
trapezoid and find its area in cm². Also find the RF scale of the globe and
its radius R.

  Partial Answer: 3.42 cm at top, **R** = 30.5 cm

1-6 Compute the great circle distances between the following pairs of cities, as
assigned. Where the coordinates are not given, obtain them from an index

in an atlas or scale them from the largest scale map of each country that you can find.

   a. Your college town and Montevideo, Uruguay
   b. Copenhagen and Melbourne
   c. Madrid (40.3°N, 3.4°W) and Capetown (33.6°S, 18.2°E)
   d. Kyoto and Baltimore

    Answer: (d) 11,120 km

1-7   What percentage of the earth's surface is in the Arctic zones (inside the Arctic and Antarctic Circles)?

    Answer: 8.3%

1-8   A portion of a circle has an arc length of 20.0 cm. If the radius is 32.0 cm, find the central angle subtended by the arc, first in radians, then in degrees. Also find the chord length.

    Partial Answer: Chord = 19.67 cm

1-9   Is the Prime Meridian a loxodrome? Is it a great circle? Is the Arctic Circle a loxodrome? Is it a great circle?

1-10 A very small square, 1.0 mm on each side, is projected as a rectangle on a certain equal-area projection. If it is 1.5 mm long, how wide is it? What are the scale factors in these two directions? There is no angular deformation between these two directions, but how much is there between the directions of the two diagonals (which were perpendicular on the globe)?

    Partial Answer: Def. = 42.07°

1-11 On a certain map of the world, the Equator is a straight standard parallel and is 47.00 cm long. If the 60th parallel also is that length, what is the scale factor along the parallel? If the projection is known to be conformal, what are the scale factors in the north-south direction at the 60th parallel and at the Equator?

    Partial Answer: Scale factor = 2.0 at 60°

REFERENCES

1. J. S. Keates, Cartographic Design and Production, John Wiley and Sons, 1973.

2. P. W. McDonnell, Cartographic Techniques (lecture notes designed to follow Introduction to Map Projections), available from the author, 1978.

3. Gurdon H. Wattles, Survey Drafting, published privately, 1977.

4. Erwin Raisz, Principles of Cartography, McGraw-Hill Book Co., 1962.

5. Robinson and Sale, Elements of Cartography, John Wiley and Sons, 1969.

2

EQUIDISTANT PROJECTIONS
WITH ONE STANDARD PARALLEL

Using the three basic projection surfaces—the cylinder, the cone, and the plane—
it is possible to generate three very simple equidistant projections. They are equidistant in the sense that all meridians are standard (the scale factor is 1.000 in
the north-south direction). In each case there is one standard parallel.

All three projections covered in this chapter are examples of an idea mentioned
in Sec. 1-3; they are not literally projected to a cylinder, cone, or plane but rather
are designed mathematically to have a desirable property. They may be thought of
as true projections on which the spacing of the parallels has been later modified to
match their spacing on the globe.

The first two in the group (the cylindrical and the conic) are not very important in themselves, but variations on them, covered in later chapters, have great
value.

CYLINDRICAL EQUIDISTANT PROJECTION

2-1 CONSTRUCTION

The cylindrical equidistant projection is also called plane chart, plate carrée,
simple cylindrical, or the cylindrical equal-spaced projection.

If a cylinder is wrapped all the way around the generating globe, touching the
Equator, the circumference of the cylinder will be the same as that of the sphere,
namely, $2\pi R$. If the whole world is to be shown on this projection, the construction is begun by drawing the Equator as a straight line of this length (the Equator
is a standard line). The meridians are standard also and are drawn as straight
vertical lines with a length of πR. Figure 2-1 shows that the resulting graticule

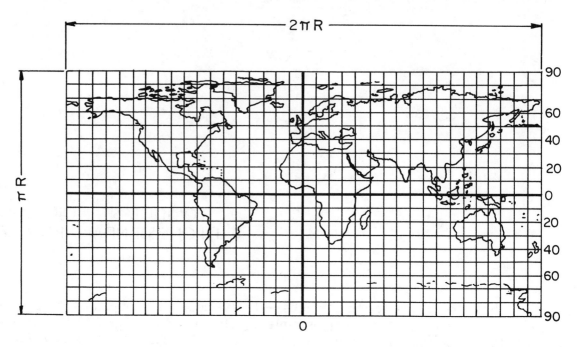

FIG 2-1 Cylindrical equidistant projection. (From Ref. 1, redrawn.)

consists of perfect squares. They are standard in their north-south dimension but, except at the Equator, they are wider than the corresponding "squares" or quadrangles on the globe. The meridians fail to converge, resulting in the north and south poles appearing as lines as long as the Equator instead of as points.

The scale factor is 1. 000 along the standard lines, by definition. It is greater than 1. 000 along the parallels. The 60th parallel, for example, has a "globe distance," or length, of $2\pi R \cos 60°$ (see Sec. 1-6) but a "map distance" equal to $2\pi R$, the same as the Equator. The scale factor is the ratio of these lengths.

$$\text{Scale factor} = \frac{\text{map distance}}{\text{globe distance}} = \frac{2\pi R}{2\pi R \cos 60°} = \sec 60° = 2.000$$

The east-west scale factor varies with sec ϕ, being 1. 000 on the Equator and infinity at the poles.

Example 2-1

Plan a cylindrical equidistant projection for a map of Africa assuming that it must fit within a 20-cm square.

Solution

Existing maps of Africa show that it is bounded, approximately, by 40°N, 35°S, 20°W, and 50°E. The width of 70° along the Equator and the height of 75° will be shown with a scale factor of 1. 000. The controlling dimension, therefore, will be north and south. The largest generating globe that could be adopted may be found by writing an expression for the length of the meridians and setting it equal to 20 cm.

$$\frac{75}{180}\pi R = 20 \text{ cm}$$

$$R = 15.28 \text{ cm}$$

This radius corresponds to the following RF scale:

$$\text{RF scale} = \frac{\text{globe radius}}{\text{earth radius}} = \frac{15.28 \text{ cm}}{637,000,000 \text{ cm}} = \frac{1}{41,690,000}$$

A smaller scale could be adopted if more generous margins are desired.

If parallels and meridians are to be plotted at 10° intervals, the dimensions of the squares will be (10/75) 20 cm = 2. 67 cm, or (10/180)π15. 28 cm = 2. 67 cm.

Table B-1 (Appendix B) may be consulted to verify that the above answers are not grossly in error (that is its only purpose).

This projection is so easy to construct that there is little need to think in terms of x and y coordinates being functions of ϕ and λ. The relationship, however, is

$$x = C\,\lambda$$
$$y = C\,\phi$$

meaning that λ and ϕ are plotted to some scale as if they were rectangular coordinates.

2-2 APPLICATIONS

The cylindrical equidistant projection is not a good choice for maps that extend very far from the Equator but is a reasonable choice if the map is limited to Indonesia or Kenya, for example. Even for the Tropics, however, there are other projections equally easy to draw that are more commonly used. The projection is included here mainly as background for the study of later chapters.

CONICAL EQUIDISTANT PROJECTION

The cylindrical equidistant projection, just discussed, was classified as cylindrical even though the spacing of the parallels was determined by the requirement that it be equidistant rather than by any actual geometric projection to a cylinder. The conical equidistant is designed in exactly the same way.

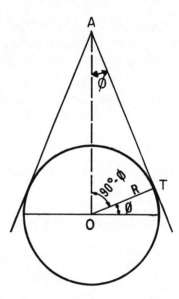

FIG 2-2 Cross-sectional view of a globe and tangent cone.

A conical equidistant projection is best suited for mapping areas in the vicinity of the standard parallel just as the cylindrical equidistant is appropriate for areas near the Equator. This projection, as well as the other conics, is generally chosen for an area lying entirely on one side of the Equator.

2-3 CONSTRUCTION

A cross-sectional view of a globe and tangent cone is shown in Fig. 2-2. The apex is at A and the point of tangency at T. The angle at the apex between the axis of the globe and the element of the cone AT is seen to be equal to the latitude of the standard parallel. In triangle ATO, the tangent of ϕ is R/AT and

$$AT = \frac{R}{\tan \phi} = R \cot \phi$$

The distance AT will be used as a compass setting r for drawing the standard parallel on the map.

The radius of the standard parallel on the globe is R cos ϕ, as shown in Fig. 1-4. Its length, of course, is $2 \pi R \cos \phi$. On the map, after the cone has been "unrolled," the parallel will have that same length but the radius used in drawing it will be AT or R cot ϕ and it will appear as less than a full circle (see Fig. 2-3). The central angle at A called L, in radians, is equal to the arc length divided by the radius.

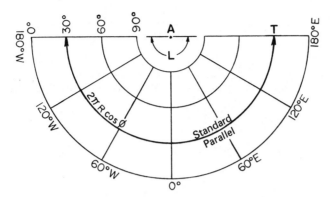

FIG 2-3 The developed cone for the Northern Hemisphere, conical equidistant projection with standard parallel at 30°.

$$L = \frac{2\pi R \cos \phi}{R \cot \phi} = 2\pi \cos \phi \, \tan \phi$$

$$= 2\pi \cos \phi \, \frac{\sin \phi}{\cos \phi} = 2\pi \, \sin \phi$$

This angle may be converted to degrees by multiplying by $360°/2\pi$.

$$L \text{ (in degrees)} = 2\pi \sin \phi \, \frac{360°}{2\pi} = 360° \sin \phi$$

The L angle is independent of scale (R was cancelled out of the expression). If $\phi = 30°$, L will be 180°, as shown in Fig. 2-3. A full 360° of longitude is shown within a semicircle. The "constant of the cone" k, defined in Sec. 1-7, is $\sin \phi$ in the case of a conical equidistant projection.

The standard parallel is divided into equal parts and the meridians are drawn as straight radial lines of standard length. As in the cylindrical equidistant, the North Pole will be a line instead of a point. For $\phi = 30°$, distance AT is 1.732R while the distance from T to the pole is $(60/180)\pi R$ or only 1.047R.

Example 2-2

Plan a graticule for mapping the Scandinavian countries within a 20 X 25 cm area using a conical equidistant projection. Show meridians and parallels at 2° intervals.

Solution

Existing maps show that the region is bounded by 4°E, 32°E, 54°N, and 72°N. The 60th parallel may be chosen as standard (three capital cities lie very close to it).

The radius for drawing the standard parallel will be

$$r = AT = R \cot 60° = 0.5774R$$

The central angle required for the 28° of longitude to be shown is

$$L = 28° k = 28° \sin 60° = 24.249°$$

The widest part of the map will be at the 54th parallel.   The radius required is

$$r_{54} = 0.5774R + \frac{6}{180} \pi R = 0.6821R$$

because all meridians are to be standard.   The width itself may be computed as a chord length (see Sec. 1-7).

$$\text{Width} = 2(0.6821R) \sin \frac{1}{2} (24.249°) = 0.2865R$$

A tentative value of R may be found by setting this term equal to the width of paper available.

$$0.2865R = 20 \text{ cm}$$
$$R = 69.81 \text{ cm}$$

The height of the map for the 18° of latitude to be shown will be

$$\frac{18}{180} \pi R = 21.93 \text{ cm}$$

plus the amount by which the 72nd parallel curves upward, if it is to be shown in full.   Clearly, the 25-cm limit is not exceeded at this scale.

Actually, a slightly larger scale is possible because part of the 54th parallel could be omitted.   (Finland, on the east, is entirely above the 60th parallel, and Denmark, the southernmost country, only extends to 8°E).

To draw this graticule with a beam compass and scale, the radius of each parallel must be calculated.   The largest radius is

$$r_{54} = 0.6821R = 47.62 \text{ cm}$$

which is within the reach of most compasses, but which involves plotting point A outside of the boundary of the map.   If the radius were beyond the reach of the compass, the parallel could be plotted by coordinates as shown in Sec. 1-7.

The central meridian will be at 18°E and may be plotted in the middle of the sheet.   Two-degree chords may be plotted along the 54th parallel, in both directions.   Straight lines may be drawn from these points to A.   For greater precision, a chord for 14° of longitude (using a central angle of 1/2 L) could be layed out along the parallel.   Finally, dividers would be used to subdivide the parallel by trial and error.

The cylindrical equidistant projection, covered in Sec. 2-1, is really only a special case of the conical equidistant in which the standard parallel is at $\phi = 0°$, radius AT equals infinity, the constant of the cone k = sin ϕ = 0, and the central angle L = 0° (the meridians being parallel). In other words, a cylinder is merely a special kind of a cone having its apex at infinity.

The conical equidistant projection is known also as the "simple conic." (The latter term, unfortunately, also has been used to describe the conic projection mentioned briefly in the opening paragraphs of Chap. 7.)

2-4 APPLICATIONS

The conical equidistant projection could reasonably be chosen for a map that covers only a few degrees of latitude such as a tourist map of the Trans-Canada Highway. The standard parallel might be 50°N in that case. Better conic projections are available, however. Five of them are discussed in the next four chapters.

POLAR AZIMUTHAL EQUIDISTANT PROJECTION

It has been pointed out that a cylinder is really a special cone with its apex at infinity. A plane which is tangent at the pole may be viewed as a special cone also. It has an altitude equal to zero and its standard parallel is at 90°N or S. (It is just a little bit flatter than a cone made tangent to 80°N, for example.) The AT distance is zero or R cot 90°. The constant of the cone will be k = sin 90° = 1.000, meaning that the central angle L, in degrees, for a full 360° of longitude will be 360°. The graticule has a fan-shaped appearance like regular conics if the fan is thought of as being wide open (see Fig. 2-4). The meridians radiate like spokes of a wheel and are separated by the same angles as they are on the globe. The projection is called the polar azimuthal equidistant. Chapter 8 discusses the non-polar, or oblique, case where the plane may be tangent to any selected point.

2-5 CONSTRUCTION

Azimuthal projections show all directions from the center without distortion. In a polar case, this means that meridians will be shown with their actual differences in longitude. If the projection is to be equidistant, all of them will be standard lines. The parallels will be equally spaced concentric circles. The opposite pole will be a large circle drawn with a radius of πR.

Example 2-3

A polar azimuthal equidistant projection of the Northern Hemisphere is to be drawn at a scale of 1:16,000,000. How large must the paper be?

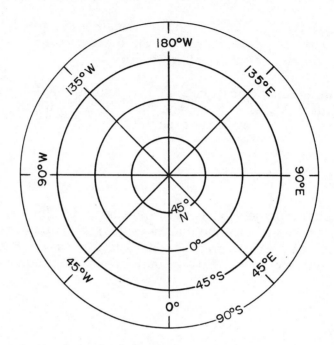

FIG 2-4 Polar azimuthal equidistant projection for the entire world.

Solution

The radius of the generating globe R is 637,000,000 cm/16,000,000 = 39.81 cm. The Equator will be a circle drawn at its true distance, which is $\pi R/2$ or 62.53 cm. The sheet will have to be 125.1 cm square, plus any desired margin.

2-6 APPLICATIONS

This projection is useful for polar exploration. The oblique case, covered in Chap. 8, has broader application because it has the azimuthal and equidistant properties with respect to any chosen point such as a radio station, airport, or seismograph rather than with respect to the north or south pole.

PROBLEMS

2-1 Determine the scale factor along the top eage of the map of Africa discussed in Example 2-1.

2-2 Plan a graticule for a cylindrical equidistant projection of Kenya assuming that it must fit within a 25-cm square. Existing maps show that your map should be bounded by 5°N, 5°S, 33°E, and 42°E. Meridians and parallels will be spaced at 1° intervals. State the RF scale and the maximum scale factor.

Partial Answer: Max. scale factor = 1.0038

2-3 Verify all of the entries in the bottom line of Table B-1 Appendix B (for a globe where R = 254 cm). Show your work.

2-4 Assume that a table similar to Table B-1 Appendix B will be set up showing only metric units. Compute the values for R = 50.0 cm or for any other assigned value.

Partial Answer: For R = 50, 1 cm = 127.4 km

2-5 Assume that a relatively small area at 60°N has been plotted on a cylindrical equidistant of the world. How many times larger, in area, will it appear than a similar feature which lies at the Equator? (For convenience, visualize it as a 100-km square.)

Partial Answer: 200 X 100 km

2-6 For the map of Scandinavia discussed in Example 2-2, find the scale factor along the 54th parallel. Also find the chord for 14° of longitude as described there.

Partial Answer: Chord = 10.06 cm

2-7 For the map referred to in the previous problem, find r_{72} and the scale factor along the 72nd parallel. Note that the map is neither conformal nor equal area because at this latitude the minimum scale factor is 1.000 and the maximum is greater than 1.000. See Secs. 1-11 and 1-12.

Partial Answer: Scale factor = 1.031

2-8 A map of Hudson Bay, in Canada, is being planned. It will use the conical equidistant or simple conic projection. The scale will be 1:8,000,000, the standard parallel will be 55°N, and the region is bounded by 50°N, 66°N, 75°W, and 95°W. Compute the width of paper required, not including any margin.

Partial Answer: r_{50} = 62.70 cm

2-9 For the map of the Northern Hemisphere in Example 2-3, find the scale factor along the 20th parallel. Would it be different for a smaller scale map?

2-10 Draw a graphic scale for Example 2-3 (an RF scale of 1:16,000,000). Make the divisions 500 and 1000 km (or miles if assigned). Show the computation.

REFERENCES

1. C. H. Deetz and O. S. Adams, Elements of Map Projection, Special Publication No. 68, U.S. Coast and Geodetic Survey (now National Geodetic Survey), Washington, D.C., 1944.

3

EQUIDISTANT PROJECTIONS
WITH TWO STANDARD PARALLELS

The cylindrical and conic projections described in Chap. 2 can be modified to have slightly smaller scale factors along all of their parallels. The two parallels having a scale factor of 1.001, for example, can be made standard, thus making the overall range of scale factors vary from 0.999 to infinity instead of from 1.000 to infinity. Scale factors then will be close to 1.000 over a larger portion of the map. This reduction of scale factors cannot be accomplished on the azimuthal equidistant projection, as will be seen.

CYLINDRICAL EQUIDISTANT
WITH TWO STANDARD PARALLELS

3-1 CONSTRUCTION

This projection is constructed in the same way as the cylindrical projection in Chap. 2 with the exception just noted, namely, that the Equator is drawn with a scale factor less than 1.000. The network of perfect squares shown in Fig. 2-1 is compressed to a smaller east-west dimension. See Fig. 3-1. As a result, this projection is referred to as the equirectangular projection by some authors [2, 3].

As a result of the east-west compression, there will be two parallels which were too long on the original projection but which are now correct.

Example 3-1

Use the cylindrical equidistant projection with two standard parallels to plan a 40 X 50-cm map of Rhode Island. Determine the minimum and maximum scale factors at the edges of the map.

30

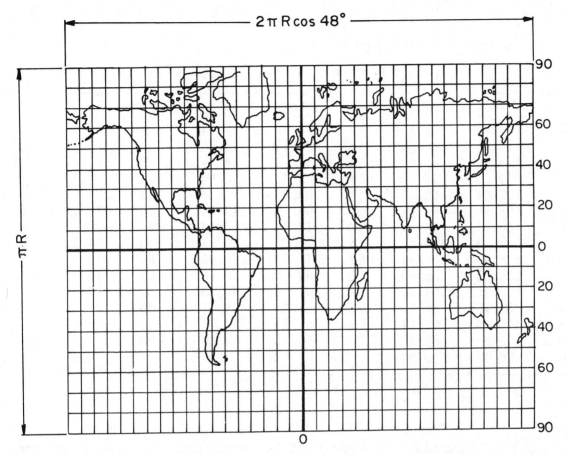

FIG 3-1 Cylindrical equidistant projection with two standard parallels (the equi-rectangular). The 48th parallels are standard in the case shown. (From Ref. 1, redrawn.)

Solution

Existing maps show that the state falls within the following coordinates: 40°07'N, 42°03'N, 71°05'W, and 71°53'W. The width of the region to be mapped is 48 minutes of longitude. The height is 56 minutes of latitude. A suitable standard parallel would be 41°30'N.

In terms of R, the width will be

$$\frac{48/60}{360} \, 2\pi R \cos 41°30' = 0.01046R$$

and the height will be

$$\frac{50/60}{360} \, 2\pi R = 0.01629R$$

TABLE 3-1

Lengths of Degrees on Selected Parallels

ϕ	km	Miles	ϕ	km	Miles	ϕ	km	Miles
0	111.32	69.17	32	94.50	58.72	62	52.40	32.56
2	111.25	69.13	34	92.39	57.41	64	48.93	30.41
4	111.05	69.00	36	90.17	56.03	66	45.41	28.22
6	110.72	68.80	38	87.84	54.58	68	41.82	25.99
8	110.24	68.50	40	85.40	53.06	70	38.19	23.73
10	109.64	68.13	42	82.85	51.48	72	34.51	21.44
12	108.90	67.67	44	80.21	49.84	74	30.78	19.13
14	108.04	67.13	46	77.47	48.14	76	27.02	16.79
16	107.04	66.51	48	74.63	46.37	78	23.22	14.43
18	105.91	65.81	50	71.70	44.55	80	19.39	12.05
20	104.65	65.03	52	68.68	42.68	82	15.54	9.66
22	103.26	64.17	54	65.58	40.75	84	11.68	7.26
24	101.75	63.23	56	62.40	38.77	86	7.79	4.84
26	100.12	62.21	58	59.14	36.74	88	3.90	2.42
28	98.36	61.12	60	55.80	34.67	90	0	0
30	96.49	59.96						

The sheet should be turned with the 50-cm dimension in the north-south direction. If the north-south dimension is controlling, R will be as follows:

$$R = \frac{50}{0.01629} = 3069 \text{ cm}$$

If the east-west dimension controls,

$$R = \frac{40}{0.01046} = 3824 \text{ cm}$$

The smaller scale must control; therefore, R will be 3069 cm. The width of the graticule will be 0.01046 X 3069 = 32.10 cm. Meridians and parallels may be drawn at an interval of 5 minutes or 10 minutes to form equal rectangles.

The scale factor at the top of the sheet will be

$$\frac{\text{Map distance}}{\text{Globe distance}} = \frac{32.10}{\frac{48/60}{360} 2\pi(3069)\cos 42°03'}$$

and therefore it is also equal to the ratio of cosines.

TABLE 3-2

Lengths of Selected Degrees on Meridians

ϕ	km	Miles	ϕ	km	Miles	ϕ	km	Miles
0-1	110.57	68.70	30-31	110.86	68.88	60-61	111.42	69.24
2-3	110.57	68.70	32-33	110.89	68.90	62-63	111.46	69.26
4-5	110.57	68.71	34-35	110.93	68.93	64-65	111.49	69.28
6-7	110.58	68.71	36-37	110.96	68.95	66-67	111.52	69.29
8-9	110.59	68.72	38-39	111.00	68.97	68-69	111.55	69.31
10-11	110.60	68.73	40-41	111.04	69.00	70-71	111.57	69.33
12-13	110.62	68.74	42-43	111.08	69.02	72-73	111.60	69.34
14-15	110.64	68.75	44-45	111.12	69.05	74-75	111.62	69.36
16-17	110.66	68.76	46-47	111.16	69.07	76-77	111.64	69.37
18-19	110.68	68.77	48-49	111.20	69.10	78-79	111.65	69.38
20-21	110.70	68.79	50-51	111.24	69.12	80-81	111.67	69.39
22-23	110.73	68.80	52-53	111.28	69.14	82-83	111.68	69.40
24-25	110.76	68.82	54-55	111.32	69.17	84-85	111.69	69.40
26-27	110.79	68.84	56-57	111.35	69.19	86-87	111.70	69.40
28-29	110.82	68.86	58-59	111.39	69.21	88-89	111.70	69.41

$$\text{Scale factor} = \frac{\cos 41°30'}{\cos 42°03'} = 1.00861$$

At the bottom edge the scale factor is

$$\frac{\cos 41°30'}{\cos 41°07'} = 0.99414$$

Distances may be scaled on this map with errors no greater than 0.86%.

Because of the large scale of the map in the preceding example, it may be advisable to refer to special tables rather than assume that the earth is a perfect sphere. Tables 3-1 and 3-2 show the lengths of a degree on every second parallel and at similar intervals along a meridian. More complete tables are available [2]. In Table 3-2 the smaller values near the Equator are caused by the greater curvature of the earth's surface.

For the map of Rhode Island in the preceding example, the RF scale was

$$\frac{3069 \text{ cm}}{637,000,000 \text{ cm}} = \frac{1}{207,560}$$

Using the tables, the "equal rectangles" may be computed for $\phi = 41°30'$. By interpolation, the width of the map (48 minutes) and the height (56 minutes) will be

$$\text{Width} = \frac{1}{207,560} (83.49 \text{ km}) \left(\frac{48}{60}\right) = 0.0003218 \text{ km}$$

$$= 32.18 \text{ cm, compared to } 32.10 \text{ in the example.}$$

$$\text{Height} = \frac{1}{207,560} (111.06 \text{ km}) \left(\frac{56}{60}\right) = 0.0004994 \text{ km}$$

$$= 49.94 \text{ cm, compared to } 50.00 \text{ in the example.}$$

The change in each overall dimension is less than a millimeter.

3-2 APPLICATIONS

The cylindrical equidistant with two standard parallels is useful for small areas away from the Equator, as demonstrated in the example. In such a case, parallels will be close to their correct lengths, meridians will be standard, and the graticule will be easy to draw.

For a map of a region on the Equator, such as the one of Kenya described in Prob. 2-2, this projection will show both standard parallels. This will give the map a better average scale than would the projection with only one standard parallel. If standard parallels are chosen so that about two-thirds of the country lies between them, the east-west scale factor will deviate from 1.000 by a minimal amount. Figure 3-2 shows a plotting of scale factor versus latitude for both cases (one and two standard parallels).

This projection is rarely used for the entire world because better projections exist. Problem 3-3, however, discusses an interesting choice of standard parallels for such a world map.

CONICAL EQUIDISTANT WITH TWO STANDARD PARALLELS

3-3 CONSTRUCTION

In the conical equidistant or simple conic described in Chap. 2, the radius called AT, used for drawing the standard parallel, was computed for an <u>actual cone</u> which was tangent to the globe. The other parallels were added mathematically, without truly projecting them. In the conical equidistant with two standard parallels there really is no cone at all. The radius for drawing each standard parallel comes directly from the decision that two parallels and all meridians should be standard.

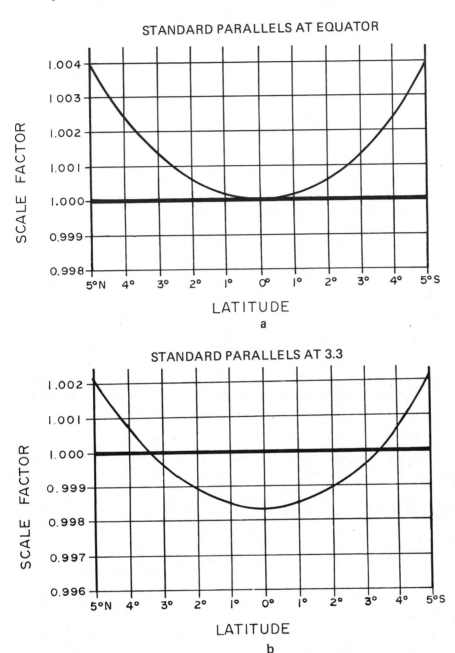

FIG 3-2 The variation of the east-west scale factor in two maps of Kenya.   The graphs represent the cylindrical equidistant projection with one standard parallel (a) and with two standard parallels (b).

In Fig. 3-3 the latitude of the northern standard parallel is called ϕ_n and that of the southern one is ϕ_s. The width of the region to be mapped, in degrees of longitude, is called $\Delta\lambda$.   The radius needed to draw the northern standard parallel is called r_n and the central angle at A is called L.   Parallel spacing d is standard.

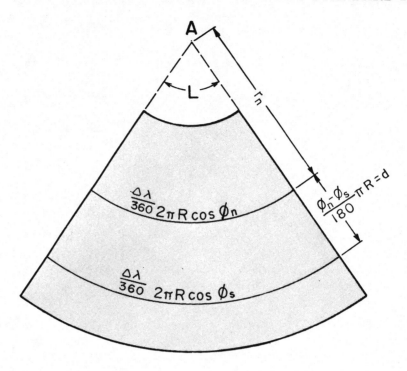

FIG 3-3 Conical equidistant projection with two standard parallels.

The angle L, in radians, is equal to either of the standard arcs divided by its radius

$$L = \frac{(\Delta\lambda/360)\, 2\pi R \cos\phi_S}{r_n + d} = \frac{(\Delta\lambda/360)\, 2\pi R \cos\phi_n}{r_n} \tag{1}$$

The two equivalent expressions for L may be used to solve for r_n. Cancelling out the common terms,

$$\frac{\cos\phi_S}{r_n + d} = \frac{\cos\phi_n}{r_n}$$

and, rearranging,

$$\cos\phi_S = \frac{(r_n + d)\cos\phi_n}{r_n} = \cos\phi_n + \frac{d\cos\phi_n}{r_n}$$

Finally,

$$r_n = \frac{d\cos\phi_n}{\cos\phi_S - \cos\phi_n} \tag{2}$$

The standard distance d is some fractional part of πR, the length of a meridian.

$$d = \frac{\phi_n - \phi_s}{180°} \pi R \tag{3}$$

Equation (3) also may be thought of in terms of radians: d is an arc distance and $\pi/180°$ merely converts $\phi_n - \phi_s$ to radians.

The constant of the cone k is the ratio of L to $\Delta\lambda$ as stated in Sec. 1-7. An expression for k may be obtained by converting L in Eq. (1) to degrees and dividing by $\Delta\lambda$.

$$k = \frac{R \cos \phi_n}{r_n}$$

Because r_n is not known initially, it is useful to modify this expression by a substitution in the denominator using Eqs. (2) and (3). The result is

$$k = \frac{180}{\pi} \left(\frac{\cos \phi_s - \cos \phi_n}{\phi_n - \phi_s} \right) \tag{4}$$

Example 3-2

Redesign the graticule for the Scandinavian countries discussed in Example 2-2 using a conical equidistant with two standard parallels. As before, the map should fit inside a 20 X 25 cm area; be bounded by 4°E, 32°E, 54°N, and 72°N; and show a graticule interval of 2°.

Solution

Generally the standard parallels are selected so that they include two-thirds of the latitudinal distance being mapped. To satisfy this rule the parallels should be at 57° and 69°, but if only even-numbered parallels and meridians are to be shown, as before, it might be more convenient to select 58° and 68°. Distance d, between them, will be kept the same as on the globe.

$$d = \frac{68° - 58°}{180°} \pi R = 0.1745R$$

The radii r_n and r_s will be

$$r_{68} = \frac{0.1745R \cos 68°}{\cos 58° - \cos 68°} = 0.4209R$$

$$r_{58} = 0.4209R + 0.1745R = 0.5954R$$

and the largest radius, 14° south of the 68th parallel, will be

$$r_{54} = 0.4209R + \frac{14}{180} \pi R = 0.6652R$$

The constant of the cone is

$$k = \frac{180}{\pi} \left(\frac{\cos 58° - \cos 68°}{68° - 58°} \right) = 0.8899$$

and the central angle L is

$$L = k\Delta\lambda = (0.8899)(28°) = 24.92°$$

Equation (1), giving L in radians, could be used instead.

The width of the map in terms of R is a chord which may be set equal to 20 cm, as was done in Example 2-2.

$$2(0.6652R) \sin \frac{24.92°}{2} = 20 \text{ cm}$$
$$R = 69.68 \text{ cm}$$

This radius is slightly smaller than the value found for the simple conic. Thus there is no need to verify that the height will be within 25 cm. The remaining steps in the construction are the same as in Chap. 2.

3-4 APPLICATIONS

While this projection is clearly superior to the conic with one standard parallel, it is neither conformal nor equal-area. Future chapters will discuss conic projections which retain the idea of two standard parallels but which also have one or the other of these properties.

This projection is used in some atlases, usually for individual countries rather than continents.

3-5 IMPOSSIBILITY OF AN AZIMUTHAL CASE

This chapter discusses cylindrical and conic equidistant projections with two standard parallels. Although it would be possible to have an azimuthal projection with two standard parallels, such as at 70°N and at the pole, it could not be equidistant. Using 70° and the pole to illustrate the difficulty, it may be recalled that the radius of the 70th parallel, if the parallel is drawn to correct length, will be R cos 70° or 0.342R. If the meridian from 70°N to 90°N is to be standard, the radius must be

$$\frac{20}{180} \pi R = 0.349R$$

There is not enough room inside of a standard parallel to draw an equidistant meridian.

PROBLEMS

PROBLEMS

3-1 Problem 2-2 involved the planning of a graticule for a cylindrical equidistant projection of Kenya to fit within a 25-cm square. The region is bounded by 5°N, 5°S, 33°E, and 42°E. The RF scale was found to be 1:4,447,000. Using this scale, compute the exact width and height of the graticule with data from Tables 3-1 and 3-2 rather than assuming that the earth is spherical. Compare your result to the answer obtained before, namely, 22.50 cm X 25.00 cm.

   Partial Answer: Height = 24.86 cm

3-2 Plan a cylindrical equidistant projection with two standard parallels for the State of Ohio using a scale of 1 cm = 20.0 km or 1 inch = 30.0 miles, as assigned. Include the area bounded by 80°W, 85°W, 38°N, and 42°N. Use the 40th parallel as the standard and use 6370 km or 3960 miles as the radius of the earth. Find (1) the RF scale and (2) the required dimensions of the sheet.

3-3 If the cylindrical equidistant projection with two standard parallels is used for the whole world, the height will be πR as shown in Fig. 3-1. If the length of the Equator is made equal to 4R the area of the whole graticule will be $(4R)(\pi R) = 4\pi R^2$. This is the same as the surface area of the generating globe of radius R. Which parallels are standard in this case? (Note that the projection is not equal area except in the overall sense.)

3-4 For the map of Scandinavia discussed in Example 3-2, calculate the scale factor along the 54th, 62nd, 64th, and 72nd parallels (it is 1.000 along the 58th and 68th, of course). Plot a diagram similar to Fig. 3-2 to show this variation graphically. (Note that the north-south scale factor is 1.000 at all of these latitudes.)

3-5 Using R = 50.8 cm or 20 inches, as assigned, compute the radii needed to draw the two standard parallels for an equidistant conic projection of the continental United States. The parallels will be 30°N and 45°N. Compute the x and y coordinates needed for plotting the 30th parallel by tangent offsets, using 95°W as the central meridian. Draw the eastern half of this parallel if assigned.

   Partial Answer: At 65°, x = 22.65 cm

3-6 Modify the expression given for r_n in Eq. (2) to make it apply to maps in the Southern Hemisphere. In this case it should be called r_s, the smaller of the two radii.

REFERENCES

1. C. H. Deetz and O. S. Adams, Elements of Map Projection, Special Publication No. 68, U.S. Coast and Geodetic Survey (now National Geodetic Survey), Washington, D.C., 1944.

2. A. H. Robinson and R. D. Sale, Elements of Cartography, 3rd Edition, John Wiley and Sons, 1969.

3. Erwin Raisz, Principles of Cartography, McGraw-Hill Book Co., 1962.

4

EQUAL-AREA PROJECTIONS
WITH STRAIGHT MERIDIANS

This chapter introduces three projections which are only slight modifications of ones covered in Chaps. 2 and 3. In each case the change made is in the spacing of the parallels. The purpose is to make the projections equal-area. The scale along each meridian, which was 1.000 in all five projections covered in previous chapters, is made to be <u>compensatory</u> instead (see Sec. 1-11).

As might be expected, the three new projections include one cylindrical, one conic, and one azimuthal.

CYLINDRICAL EQUAL-AREA PROJECTION

4-1 CONSTRUCTION

The cylindrical, or rectangular, equal-area projection resembles the cylindrical equidistant in several ways. The Equator is the only standard parallel and the meridians are perpendicular to it. The parallels have the same length as in the cylindrical equidistant and the east-west scale factor again varies with sec ϕ.

Along the 60th parallel the scale factor is 2.000. The compensatory scale factor, along the meridian, must be 0.500. (The product of the minimum and maximum scale factors at any point must be 1.000 if the projection is to be equal-area.) The reduced north-south scale factors, such as 0.500 at $\phi = 60°$, cause a compression of the perfect squares found on the equidistant. See Fig. 4-1b.

Section 1-6 and Fig. 1-5 point out that the surface area of a sphere, or globe, happens to be equal to that of a surrounding cylinder of the same height (2R). Therefore this projection may be formed by projecting or extending the planes of the various parallels of a globe until they intersect a surrounding cylinder. (Of

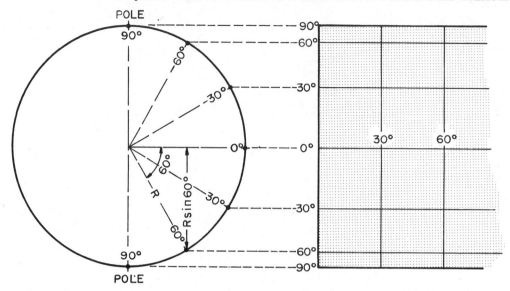

FIG 4-1a Cylindrical equal-area projection and its generating globe.

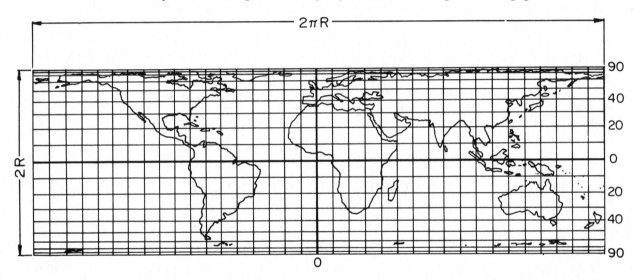

FIG 4-1b World map on a cylindrical equal-area projection.   (From Ref. 1.)

the projections covered thus far, this is the only one which is literally projected.)
Figure 4-1a shows this construction, using a cross-sectional view of a globe.   If
the projection is plotted by calculation rather than by graphic construction, the
distance from the Equator to any parallel is R sin ϕ.

Example 4-1

Plan a map of Africa to fit inside a 20-cm square using the cylindrical equal-
area projection. The region is bounded by 40°N, 35°S, 20°W, and 50°E. (This
problem is similar to Example 2-1, in which the cylindrical equidistant was
used.)

Solution

The height of the map from 40°N to 35°S will be

Height = R sin 40° + R sin 35° = 1.216R

The width, from 20°W to 50°E, will be

$$\frac{70}{360}\, 2\pi R = \frac{70}{180}\, \pi R = 1.222R$$

The map will be slightly wider than it is high and R may be found by fitting this width into the 20-cm limit.

1.222R = 20 cm

   R = 16.37 cm

If the parallels and meridians are to be plotted at 10° intervals, the spacing of meridians will be

$$\frac{10}{70}\, 20 \text{ cm} = 2.86 \text{ cm}$$

and the parallels will be above and below the Equator at distances equal to R sin 10°, R sin 20°, R sin 30°, and R sin 40°.

4-2 APPLICATIONS

The cylindrical equal-area projection is rarely used, but is a reasonable choice for areas which straddle the Equator. Generally, equal-area maps show large portions of the earth, or all of it, in which case there are some other projections that are better. They show less extreme distortion or compression away from the Equator.

Generally, equal-area maps are used to show distributions, such as population density. If a small dot is used to represent a million people it would be important that each country appear in its proper relative size. (Otherwise a dense collection of dots might not represent a dense population.)

ALBERS PROJECTION

4-3 CONSTRUCTION

From the description of the conical equidistant projection with two standard parallels (Chap. 3), it can be seen that such a map shows less area between the standard

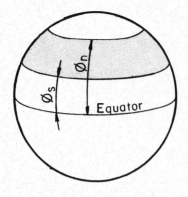 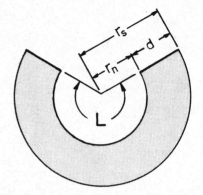

FIG 4-2   Zone on the globe and the circular strip on a map which must be equal on Albers projection.

parallels than would an equal-area map. The scale factor along all meridians was 1. 000, and that along the parallels (in that portion of the map) was less than 1. 000 (see Prob. 3-4). The product of these values also is less than 1. 000, indicating a reduction in area. The Albers projection, in order to be equal-area, allows the scale factor along the meridians to be greater than 1. 000 to compensate for the smaller east-west scale factors. The two standard parallels thus are farther apart in the equal-area map than in the equidistant map. The projection is named for Dr. H. C. Albers of Germany, who invented it in 1805 [1]. The following discussion is based on Ref. 2.

To calculate the increased distance d between the standard parallels, the correct area of the zone (on the globe) is set equal to that of the circular strip to be drawn on the map, with d as an unknown. In Fig. 4-2, d is shown to be $r_s - r_n$.

The surface area between parallels ϕ_n and ϕ_s on the globe was shown in Sec. 1-6 to be

$$2\pi R \,(R \sin \phi_n - R \sin \phi_s) = 2\pi R^2 \,(\sin \phi_n - \sin \phi_s)$$

where ϕ_n is the larger of the two latitudes and ϕ_s is the smaller. (The n and s are appropriate letters in the case of the Northern Hemisphere.)

On the map projection, the circular strip which corresponds to 360° of longitude will have an unknown central angle L, as shown in Fig. 4-2. The fraction of a full circle represented by L is the constant of the cone k, as discussed in earlier chapters. If L is in degrees, $k = L/360°$. The area of the strip will be the difference of two sectors.

$$k\pi r_s^2 - k\pi r_n^2 = k\pi (r_s^2 - r_n^2)$$

These two areas, one on the sphere and one on the map, are to be equal.

$$k\pi (r_s^2 - r_n^2) = 2\pi R^2 (\sin \phi_n - \sin \phi_s)$$

The lengths of the standard parallels must be

$$2 \pi R \cos \phi_n \quad \text{and} \quad 2 \pi R \cos \phi_s$$

while on the map these same distances will be equal to

$$k(2 \pi r_n) \quad \text{and} \quad k(2 \pi r_s).$$

Thus

$$r_n = \frac{R \cos \phi_n}{k} \quad \text{and} \quad r_s = \frac{R \cos \phi_s}{k} \tag{1}$$

These expressions may be substituted into the equal-area equation

$$k \pi \left(\frac{R^2 \cos^2 \phi_s}{k^2} - \frac{R^2 \cos^2 \phi_n}{k^2} \right) = 2 \pi R^2 (\sin \phi_n - \sin \phi_s)$$

which simplifies to

$$\frac{\cos^2 \phi_s - \cos^2 \phi_n}{k} = 2(\sin \phi_n - \sin \phi_s)$$

from which

$$k = \frac{\cos^2 \phi_s - \cos^2 \phi_n}{2(\sin \phi_n - \sin \phi_s)}$$

Using trigonometric identities in the numerator, this reduces to

$$k = \frac{1}{2}(\sin \phi_n + \sin \phi_s) \tag{2}$$

which may be substituted into the expressions for r_n and r_s.

With k known, it is possible to obtain the radius r_1 of any other parallel ϕ_1. The area on the map between this parallel and the northern standard parallel is

$$k \pi r_1^2 - k \pi r_n^2$$

This must be set equal to the corresponding area on the globe. The equality takes the same form as for the areas between the standard parallels, but both k and r_n are no longer unknowns.

$$k \pi r_1^2 - k \pi r_n^2 = 2 \pi R^2 (\sin \phi_n - \sin \phi_1)$$

From this,

$$kr_1^2 = kr_n^2 + 2R^2 (\sin \phi_n - \sin \phi_1)$$

$$r_1 = \left(\frac{kr_n^2 + 2R^2 (\sin \phi_n - \sin \phi_1)}{k} \right)^{1/2} \tag{3}$$

Example 4-2

Plan a graticule for an Albers projection of Scandinavia using $R = 100$ cm, $\phi_n = 68°$, and $\phi_s = 58°$. Use a programmable pocket calculator to generate x and y coordinates for each 2° on each parallel. As the origin for the coordinates, use $\phi = 54°N$, $\lambda = 18°E$. Note that this graticule could be plotted by beam compass and scale instead, especially if the radii were not too large.

Solution

The region to be mapped is bounded by 4°E, 32°E, 54°N, and 72°N. Because the graticule will be symmetrical about the central meridian (18°E), it is only necessary to calculate coordinates for the eastern half. In this half there will be eight points along each of 10 parallels. The points may be plotted as each one is calculated, using a coordinatograph, flat-bed plotter, or drafting machine, or by using a sheet with a preprinted grid. (Sheets are available with a no-print grid that is visible to the eye but which will not appear on an Ozalid copy.)

Several calculations will be made before programming a pocket calculator to generate the 80 sets of coordinates. First, the constant of the cone is

$$k = \frac{1}{2}(\sin 68° + \sin 58°) = 0.88762$$

The longitudinal coverage of the map ($\Delta \lambda$) is 28° and the central angle on the graticule will be

$$L = 28°k = 24.853°$$

For each 2° interval of longitude, the central angle L' will be

$$L' = 2°k = 1.7752°$$

The radii of the upper standard parallel and the 54th parallel will be

$$r_n = r_{68} = \frac{100 \cos 68°}{0.88762} = 42.20 \text{ cm}$$

$$r_{54} = \left(\frac{0.88762(42.20)^2 + 2(100)^2 (\sin 68° - \sin 54)}{0.88762} \right)^{1/2}$$

$$= 66.66 \text{ cm}$$

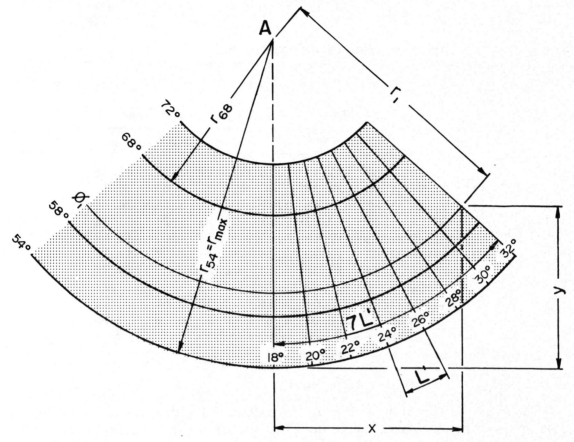

FIG 4-3 Albers projection for Scandinavia (not to scale) showing standard parallels at 58° and 68° and coordinates for plotting a point on another parallel ϕ_1.

The program will begin by calculating the radius r_1 of a particular parallel. It will use that radius and a central angle such as 0°, L', 2L', 3L', 4L', etc., as polar coordinates and then display rectangular coordinates. Within the program, the y coordinate based on the apex A as an origin will be transformed to one with the desired origin by subtracting it from r_{54}. See Fig. 4-3.

The calculator used in this example is an HP-25. Five constants will be stored as follows:

k   in STO 1

kr_n^2 in STO 2

$2R^2$ in STO 3

$\sin \phi_n$ in STO 4

r_{54} or r_{max} in STO 5

TABLE 4-1

Plotting Table for Albers Projection of Scandinavia

$\Delta\lambda$: Central Angle:	0° 0.00°		2° 1.7752°		4° 3.5504°		6° 5.3256°		8° etc.	
ϕ	x	y	x	y	x	y	x	y	x	y
72°	0	31.40	1.09	31.42	2.18	31.47	3.27	31.56		
70°	0	27.94	1.20	27.96	2.40	28.02	3.59	28.11		
68°	0	24.46	1.31	24.48	2.61	24.54	3.92	24.64		
66°	0	20.96	1.42	20.99	2.83	21.05	4.24	21.16		
64°	0	17.46	1.52	17.48	3.05	17.56	4.57	17.67		
62°	0	13.96	1.63	13.98	3.26	14.06	4.89	14.18		
60°	0	10.46	1.74	10.48	3.48	10.56	5.22	10.70		
58°	0	6.96	1.85	6.99	3.70	7.08	5.54	7.22		
56°	0	3.48	1.96	3.51	3.91	3.60	5.86	3.75		
54°	0	0.00	2.06	0.03	4.13	0.13	6.19	0.29		

The program is shown in Sec. A-1. To calculate the coordinates of each point one must store the central angle in STO 0 and put ϕ_1 in the x register. The program will generate and display the y coordinate. The x-y interchange key is then used to display the x coordinate.

If the coordinates are tabulated before being plotted, the arrangement will be as shown in Table 4-1. It is convenient to work down the columns, storing a new central angle each time. If desired, the program could be made to perform some of the preliminary steps in addition to the repetitive ones.

4-4 APPLICATIONS

The Albers projection is used on most pages of the National Atlas of the United States where the continental United States is shown. The equal-area property is useful for the problems of geography and, for the region involved, the projection is not far from being conformal. If an Albers projection of the contiguous states is drawn at a scale of 1:5,000,000, it will be 1 m in width. It will differ from a conformal conic projection (Chap. 6) by only 3.6 mm at the corners [1].

The greatest scale error on such an Albers map is only 1-1/4%. The polyconic (Chap. 5) has been used in the past but has a maximum scale error of 7%.

The U.S. Geological Survey uses standard parallels at 29°30' and 45°30' for the National Atlas, and special tables for plotting the graticule with beam compass and scale are available [1]. (These tables use the spheroid as the datum.)

Another equal-area conic (the Bonne projection) will be covered in Chap. 5. It is more difficult to draw because it has curved meridians.

POLAR AZIMUTHAL EQUAL-AREA PROJECTION

4-5 CONSTRUCTION

This projection resembles the polar azimuthal equidistant (Chap. 2), but different radii are used to draw the parallels. Each parallel, such as the Arctic Circle, is drawn so that it contains the same area on the flat paper as exists poleward of the parallel on the generating globe.

For any parallel, the area on the map will be simply πr^2 and the surface area between the pole and the parallel on the globe will equal that of the strip of similar height on the surrounding cylinder (see Sec. 1-6). That surface area is

$$2\pi R(R - R \sin \phi) = 2\pi R^2(1 - \sin \phi)$$

or

$$2\pi R^2[1 - \cos (90 - \phi)]$$

These two areas may be set equal to each other and solved for r, the unknown radius.

$$\pi r^2 = 2\pi R^2[1 - \cos (90 - \phi)]$$

$$r = R(2[1 - \cos (90 - \phi)])^{1/2}$$

$$= R\left(2\left[2 \sin^2 \left(\frac{90 - \phi}{2}\right)\right]\right)^{1/2}$$

$$= 2R \sin\left(\frac{90 - \phi}{2}\right) \tag{4}$$

This equation may be solved to obtain r for each parallel to be plotted.

It is also possible to find each required radius by a graphic construction. The expression for r is the same as that for the length of the chord shown in Fig. 4-4. It is the chord in the globe from the pole to the parallel.

Example 4-3

Tabulate the radii required for plotting a polar azimuthal equal-area projection covering the area north of 15°S. The size of the sheet is limited to 40 × 40 cm.

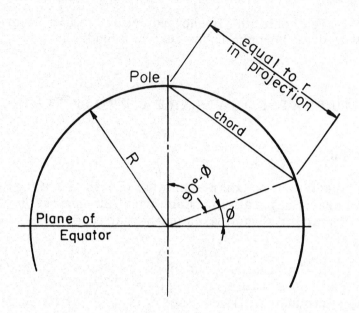

FIG 4-4 Graphic construction for polar azimuthal equal-area projection.

Solution

In the expression for r, the latitude ϕ must have its proper sign. The map extends to 15°S or -15°. The largest possible scale may be obtained by finding R as follows:

$$r_{max} = 20 \text{ cm} = 2R \sin\left(\frac{90° + 15°}{2}\right)$$

$$R = \frac{20 \text{ cm}}{2 \sin 52.5°} = 12.60 \text{ cm}$$

The radius for plotting the Equator is

$$r_0 = 2(12.60) \sin\frac{90° - 0°}{2} = 17.82 \text{ cm}$$

Similarly,

$$r_{15} = 2(12.60) \sin 37.5° = 15.34 \text{ cm}$$

$$r_{30} = 2(12.60) \sin 30° \quad = 12.60 \text{ cm}$$

and the radii for ϕ = 45°, 60°, and 75° are found in the same way.

4-6 APPLICATIONS

A hemisphere map on this projection, centered on the North Pole, is shown in Fig. 4-5.

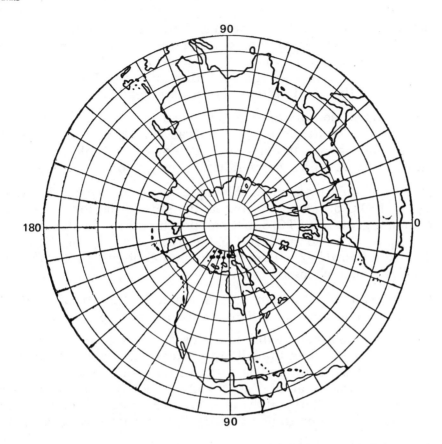

FIG 4-5 A polar azimuthal equal-area projection of a hemisphere. (From Ref. 1.)

Chapter 8 discusses the oblique case of this projection (centered on some point of interest other than the pole). Generally, the polar case is less useful, but it appears in Goode's School Atlas, for example, where hemispheres are shown with climatic themes (temperatures in January and July, etc.).

This projection is one of several credited to J. H. Lambert (the cylindrical equal-area is another). Lambert (1728-1777) made many great contributions to the study of map projections and introduced the hyperbolic functions to trigonometry [1].

PROBLEMS

4-1 For the map of Africa in Example 4-1, find the distance between the 29th and 31st parallels. (This involves first calculating their separate distances from the Equator.) Divide this "map distance" by the "globe distance" between them. Multiply this approximate scale factor by the east-west scale factor on the 30th parallel to verify that the projection is equal-area. Show your work. Make a sketch of an ellipse similar to that of Fig. 1-11 using the scale factors for the semimajor and semiminor axes.

4-2   For the cylindrical equidistant map of Kenya in Prob. 2-2, R was found to be 143.24 cm and the height of the map was 25.00 cm. Find the height of a cylindrical equal-area map using the same scale. Would the width be any different than it was using the equidistant projection?

4-3   Modify the program given for an Albers projection to make it calculate and store r_{max} in STO 5. Start by storing the sine of the lowest parallel (sin 54° in the example given) in STO 6 and storing the other constants as in the existing program. The computation and storing of r_{max} does not require all of the "stack" of an HP-25, thus the original contents of X (which is φ_1) can be rolled down when needed.

4-4   Adapt the program given for an Albers projection to a pocket calculator which uses algebraic logic instead of RPN (which the HP-25 uses).

4-5   For a map of the continental United States compute part of a plotting table similar to Table 4-1. Use the Albers projection, standard parallels at 30° and 45°, and R = 50.8 cm. (The data are the same as in Prob. 3-5.) Using an origin of coordinates at φ = 25°N and λ = 95°W, calculate x and y for the points where the 25th, 30th, and 35th parallels meet the meridian at 75°W. (There is no need to use the HP-25 program for so few points.)

       <u>Partial Answer:</u> x for 35° = 14.31 cm

4-6   Calculate x and y values for Table 4-1 corresponding to $\Delta\lambda$ = 10° and φ = 60°.

       <u>Partial Answer:</u> y = 11.13 cm

4-7   The first column of Table 4-1 shows that the 70th and 72nd parallels will be 3.46 cm apart. A more exact value is 3.462. Use this distance to calculate the average north-south scale factor in this part of the map. Also calculate the scale factor along the 71st parallel. Does it appear that the projection may be equal-area? Discuss this.

4-8   In Prob. 4-5, would it be convenient to use a beam compass instead of coordinates? What compass setting would be required for the 25th parallel? Also consider a larger scale map where R is twice as large (101.6 cm).

4-9   Find the three radii not computed in Example 4-3.

       <u>Partial Answer:</u> 3.29 cm for 75°

4-10   Verify that the polar azimuthal equal-area projection is indeed equal-area. Do this by calculating the scale factor along the Equator (map distance ÷ globe distance) and then, approximately, in the radial direction at the same latitude. For the radial scale factor, obtain the map distance between 1°S and 1°N, and the globe distance for 2° of latitude. (Note that the solution of this problem is independent of scale; R need not be known.)

REFERENCES

1. C. H. Deetz and O. S. Adams, Elements of Map Projection, Special Publica-
    tion No. 68, U.S. Coast and Geodetic Survey (now National Geodetic Survey),
    Washington, D.C., 1944.

2. J. A. Steers, An Introduction to the Study of Map Projections, 14th Edition,
    University of London Press, 1965.

5

PROJECTIONS WITH ALL PARALLELS STANDARD

All five of the projections covered in Chaps. 2 and 3 had scale factors of 1.000 along all meridians. They were equidistant in the north-south direction. This chapter introduces four projections that are equidistant in the east-west direction (all parallels are standard).

The cylindrical and conic projections of this type are designed mathematically rather than being actual geometric projections. The azimuthal one in this group is a true projection, only the second out of the 12 projections introduced up to that point.

SINUSOIDAL PROJECTION

5-1 CONSTRUCTION

To construct a sinusoidal projection of the whole world, one simply draws a standard central meridian and then constructs straight horizontal standard parallels at equal intervals (see Fig. 5-1). Because the lengths of the parallels vary with $\cos \phi$, the meridians are sine waves (or cosine waves), hence the name <u>sinusoidal</u>.

Section 1-5 pointed out that a small quadrangle on a globe, measuring 1° on a side, is very nearly a trapezoid. If it is considered a trapezoid, its area is equal to the average width times the height. In the sinusoidal projection, such a trapezoid will be distorted as shown in Fig. 5-2, but it will have the same widths and height and hence the same area. Thus the projection is equal area.

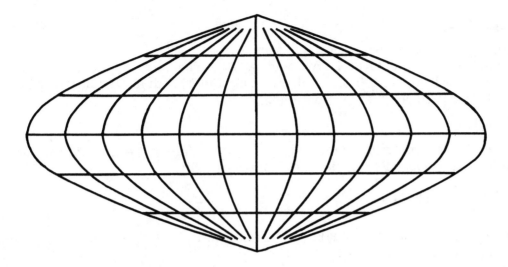

FIG 5-1 Sinusoidal projection for the entire world.   (Drawn by Glenn Watson.)
See also Fig. 1-15.

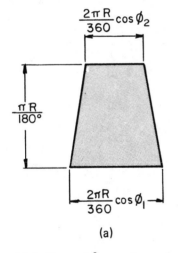

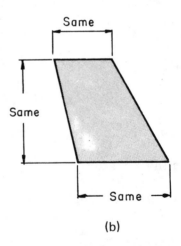

(a)   (b)

FIG 5-2 At left, a 1° quadrangle on a globe (the sides are shown as straight lines).
At right, the same quadrangle on a sinusoidal projection. The area is unchanged
because the height and two bases are unchanged. The use of straight lines would
be more nearly correct if only 1 second of latitude and longitude were used.

Example 5-1

Plan a sinusoidal projection for Africa to fit a 20-cm square. It should be bounded by 40°N, 35°S, 20°W, and 50°E. (The same specifications were used for Examples 2-1 and 4-1 involving the cylindrical equidistant and cylindrical equal-area projections.)

Solution

In all three projections mentioned, the width (along the Equator) is the same as on the globe. For Africa, 70° in width, this is

$$\frac{70}{360} 2\pi R = 1.2217R$$

The height of the sinusoidal, like that of the cylindrical equidistant, is standard also. It is, for Africa,

$$\frac{75}{180} \pi R = 1.3090R$$

To fit the map into a 20-cm square, the height will control, as it did in the cylindrical equidistant, and R will be the same, that is, 15.28 cm. The spacing of the parallels also will be the same, namely, 10/75 times 20 cm or 2.67 cm (for 10°), but their lengths will be different. Using 20°E for the central meridian, the westernmost meridian will be at $\Delta\lambda = 40°$ and may be plotted at the following distances:

For $\varphi = 40°$, $x = (40/360)\, 2\pi R \cos 40° = \ 8.17$ cm

For $\varphi = 30°$, $x = (40/360)\, 2\pi R \cos 30° = \ 9.24$ cm

For $\varphi = 20°$, $x = (40/360)\, 2\pi R \cos 20° = 10.02$ cm

For $\varphi = 10°$, $x = (40/360)\, 2\pi R \cos 10° = 10.51$ cm

For the Equator, $x = (40/360)\, 2\pi R = 10.67$ cm

The other meridians cross the parallels at equal intervals.

5-2 APPLICATIONS

The sinusoidal projection was used as early as 1606 without being given a name [1]. It became known as the Sanson-Flamsteed projection after it was used by Nicolas Sanson in 1650 and John Flamsteed in 1729. Still later it was given the name used here.

Rand McNally's world atlas uses the sinusoidal projection for Africa exactly as in the example just given. Even the scale used is very nearly the same. Because the page is not square they were able to extend the map northward to include Europe. The projection also was used for South America.

When all or most of the world is shown on a sinusoidal projection, shapes will be badly distorted as was indicated in Fig. 1-15. To reduce this problem it is common to interrupt the map in the oceans and recenter each continent on its own central meridian. In Fig. 5-3 this was done to some extent but only three central meridians were used. If the purpose of the map is to show ocean shipping, the interruptions may be placed in the land masses and the central meridians in the oceans.

The only justification for classifying the sinusoidal as cylindrical is the fact that the parallels are horizontal on the sheet as in cylindrical projections. Sometimes it and similar projections are called "pseudocylindrical" (see Mollweide projection, Chap. 9).

Of the projections covered to this point, the sinusoidal is the only one which can show both poles as points instead of as lines. It is one of several good equal-area projections suitable for a region extending on both sides of the Equator (conics are better for a region entirely in the Northern or Southern Hemisphere).

BONNE PROJECTION

5-3 CONSTRUCTION

The sinusoidal projection is really a special case of the Bonne projection. In the Bonne, a central parallel is chosen and constructed as in the conical equidistant (Sec. 2-3). A central meridian is drawn vertically with standard length and the remaining parallels are added at correct spacing and with standard length. The sinusoidal case is that in which the Equator is chosen as the central parallel and is drawn with an infinite radius (making it a straight line). The apex of its cone is at infinity. Figure 5-4 shows the relationship of the sinusoidal and Bonne projections to the cylindrical and conical equidistant projections covered in Chap. 2.

If a hemisphere is shown on a Bonne projection, as in Fig. 5-5, the resemblance to a sinusoidal projection is evident. In practice it is not normally used for that much of the earth.

Example 5-2

Plan a Bonne projection for Alaska using R = 80 cm. Adopt $\varphi = 60°$N as the central parallel and $\lambda = 160°$W as the central meridian.

Solution

See Fig. 5-6. The radius for drawing the central parallel is R cot φ. This was shown in Fig. 2-2, where it was called AT.

r = AT = 80 cot 60° = 46.19 cm

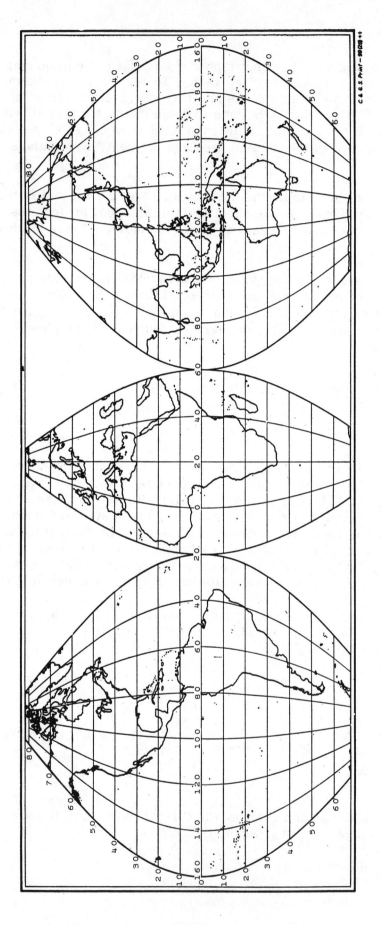

FIG 5-3 Interrupted sinusoidal projection. Three central meridians are used in this example and the polar regions are omitted. (From Ref. 1.)

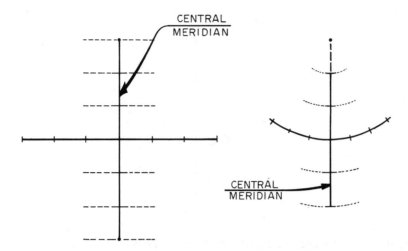

FIG 5-4 Until the meridians are added, the construction of sinusoidal and Bonne projections is identical to that of the cylindrical and conical equidistant projections. In the latter two projections the meridians are constructed as straight lines perpendicular to the standard parallel. In the sinusoidal and Bonne projections, the meridians are <u>curved</u> lines constructed by laying out all parallels as standard lines.

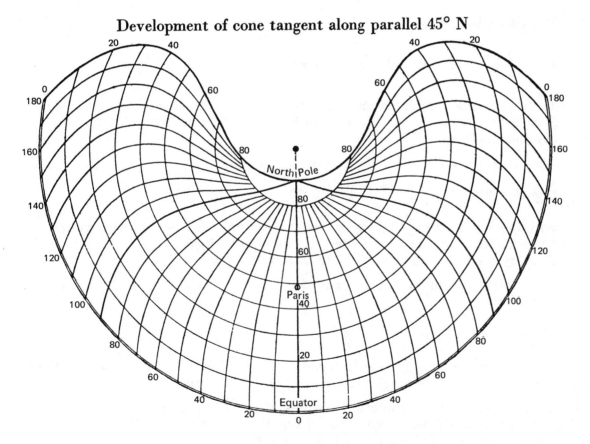

FIG 5-5 Bonne projection of a hemisphere. Because of the curved meridians, the Bonne is sometimes classified as pseudoconic rather than conic. (From Ref. 1.)

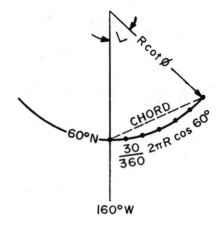

FIG 5-6 First steps in constructing a Bonne projection for Alaska.   See Example 5-2.

If the map extends 30° each way from the central meridian, this parallel will have an arc length on each side equal to

$$\frac{30}{360}\, 2\pi R \cos 60° = 20.94 \text{ cm}$$

The central angle on each side of the meridian is

$$L = \frac{20.94}{46.19} = 0.4533 \text{ radians or } 25.97°$$

This angle also may be found using k = sin φ as the constant of the cone (see Sec. 2-3). Thus L = kΔλ = 30° sin 60° = 25.98°, the difference being solely due to round-off.

The chord distance out to the end of the parallel is

$$\text{Chord} = 2(46.19) \sin\left(\frac{25.97}{2}\right) = 20.76 \text{ cm}$$

With these data it is possible to plot the central parallel and its termination 30° each way from center. Each half is then divided into six equal parts (by dividers, for example) if a 5° interval is desired.

The central meridian extends from 50°N to 75°N and is a standard line. Its length is

$$\frac{25}{180}\, \pi R = 34.91 \text{ cm}$$

and the parallels cross it at five equal intervals of 6.98 cm. They may be drawn as concentric circles using the beam compass.

As in the sinusoidal projection, the meridians cut each parallel at standard distances and will be curved lines. For the 55th parallel, the arc distance corresponding to 30° each way from center is

$$\frac{30}{360} \, 2\pi R \cos 55° \; = \; 24.03 \text{ cm}$$

and the central angle is

$$L_{55} \; = \; \frac{24.03}{46.19 + 6.98} \; = \; 0.4519 \text{ radians or } 25.89°$$

(Note that it is <u>not</u> possible to obtain k for this parallel from the relationship in a conical equidistant where it was sin ϕ.)

The chord distance to each end of the parallel is

$$\text{Chord} = 2(46.19 + 6.98) \sin\left(\frac{25.89}{2}\right) = 23.82 \text{ cm}$$

The two halves of the 55th parallel are divided into six equal parts as was done on the central parallel.

After each parallel has been constructed in this manner, the curved meridians are drawn through the proper points on each parallel. A flexible curve may be used as a drawing aid.

5-4 APPLICATIONS

The Bonne projection is equal-area, the reason being the same as for the sinusoidal (see Fig. 5-2). It is a good choice for square-shaped regions that are located on one side of the Equator such as France. In fact, it was used there beginning in 1803 for a set of topographic maps [2]. An individual sheet located some distance to the east of the central meridian has all meridians curving toward the left and all parallels curving up to the right. The Times Atlas of the World uses the Bonne for a map of Australia and New Zealand and for one of Europe. It should be noted that the Bonne and Albers projections are both equal-area conics. An apparent difference is that the Albers has straight meridians.

The Bonne projection has been used for a map of the whole world with the central parallel at 15°N.

The projection is named for Rigobert Bonne (1927-1795) but dates back before his time [3].

POLYCONIC PROJECTION

5-5 CONSTRUCTION

In the polyconic projection, each parallel is separately treated as if it were the standard parallel of a conical equidistant projection (Sec. 2-3). Each has its own radius, R cot ϕ, called AT in Chap. 2.

To begin the map, a central meridian is drawn as a standard line. The locations of the parallels are marked at equal intervals. The center from which to draw each circular arc is located on the extension of the central meridian. Each center is plotted at a distance of R cot ϕ above the point where the parallel is to cross the meridian. The result is shown in Fig. 5-7. Each parallel is standard. The effect is, as the name implies, a projection of "many cones." There is a separate constant of the cone for each parallel. It is equal to sin ϕ as in the conical equidistant. Thus the central angle L, if an entire hemisphere is to be shown, is

$$L = 360° \sin \phi$$

in each case. Each parallel, being standard, is divided at equal intervals to locate the meridian crossings. The meridians are drawn with a flexible curve through the points on each parallel. As shown in the figure, the Equator appears as a straight line, its cone being a cylinder and its radius being infinity (R cot 0°).

Obviously, the hemisphere appears oddly shaped in Fig. 5-7. The projection, however, is never used for such a large region.

Usually, the polyconic is plotted from special tables rather than with a compass as implied here. One reason is that it is used primarily on large-scale maps where the curves are very flat and beam compasses would be impractical. Also, for use at such scales, the tables are based on a spheroid rather than a sphere [4].

5-6 APPLICATIONS

For many years, the polyconic projection was used for the topographic map series produced by the U.S. Geological Survey (the so-called quad sheets covering 7-1/2 minute quadrangles at a scale of 1:24,000). Each sheet had its own central meridian. Adjacent sheets could be fitted together quite well. The fit was perfect at the tops and bottoms which, being parallels, were drawn with the same radius on both sheets. Along the sides, the fit was not quite perfect because the meridians curved in opposite directions. The Lambert conformal conic and transverse Mercator projections (state coordinates) have been used since 1953, although the marginal information continued to say "polyconic" for many years.

In a map of the United States, the stretching of the meridians, so apparent in Fig. 5-7, amounts to as much as 7% along the east and west coasts. An Albers projection, by contrast, has a maximum scale error of about 1-1/4% [1].

For many applications, the polyconic projection is being replaced by projections that are either conformal or equal area, but it appears in many atlases. The National Geographic Society uses it for the British Isles and Japan as well as for a map of Civil War battlefields. Rand McNally and Company uses it for India, Indonesia and the Philippines, China, and Alaska.

For a small area, not extending more than 3° from a central point, this projection is almost identical to a Bonne projection (Sec. 5-3). The difference is smaller than the plotting errors and paper shrinkage that would occur.

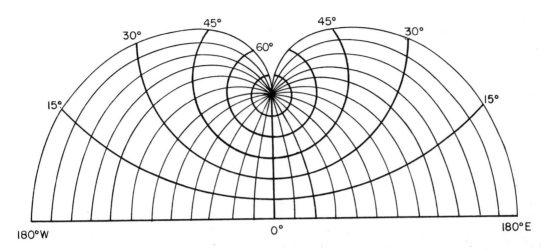

FIG 5-7 Polyconic projection of a hemisphere. Because of the curved meridians, the polyconic, like the Bonne, is sometimes classified as pseudoconic.

The projection was devised by Ferdinand Hassler of the U. S. Coast Survey (now the National Geodetic Survey) in 1820 [3]. Modifications have been developed by others [1].

POLAR ORTHOGRAPHIC PROJECTION

The word <u>orthographic</u> is familiar to anyone who has studied engineering or architectural drawing. House plans show several orthographic projections such as front and side views or "elevations" of the proposed house. The floor plan itself is an orthographic top view.

An orthographic projection is one in which the projecting rays are perpendicular to the plane of projection. Any part of the object that is parallel to the plane of projection will appear in its proper shape and correct scale. One can scale the dimensions of a door or window on the front view of a house, for example.

An orthographic view of a globe looks very much like a photograph of the globe taken from a distance. The further away one is when looking at a globe the more nearly orthographic will be the view. An orthographic view is a view from infinity and thus lacks perspective. It can show a full hemisphere, but no more. Figure 1-2 is an orthographic projection centered on the North Atlantic Ocean. Our view of the moon is nearly orthographic.

5-7 CONSTRUCTION

Figure 5-8 shows the formation of an <u>oblique</u> orthographic projection such as the one just referred to (Fig. 1-2). A <u>polar</u> case is formed in the same way but

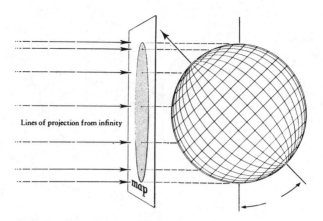

FIG 5-8 The orthographic projection of a globe to a plane. The plane shown will contain an oblique projection rather than the polar case discussed in this chapter. The drawing of the globe used in the figure itself is an equatorial orthographic. (From <u>Maps and How to Understand Them</u> by Consolidated Vultee Aircraft Corp., courtesy of General Dynamics.)

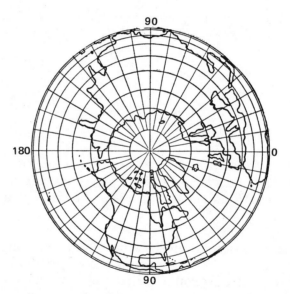

FIG 5-9 Polar orthographic projection.   (From Ref. 1.)

"looks" at the globe from directly above the north or south pole. Figure 5-9 shows the northern case. All parallels of latitude appear with their proper diameters; each is drawn with a radius r = R cos ϕ and is therefore a standard parallel. This is also evident in a graphic construction. See Fig. 5-10.

Meridians radiate at equal angles exactly as in the other polar azimuthal projections (Chaps. 2 and 4).

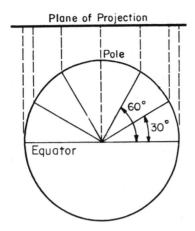

FIG 5-10 Graphic construction of the polar orthographic projection.

5-8 APPLICATIONS

Polar as well as oblique orthographic projections are useful because of their globe-like appearance (they look more like a globe than a map). They are used in art and illustration work to convey a realistic view and sometimes are shown without the graticule to give almost an astronaut's view of the earth.

PROBLEMS

5-1 Design the largest possible sinusoidal projection of South America for a 43 X 55 cm area of paper. The region is bounded by 90°W, 30°W, 60°S, and 20°N. Assume a 10° interval in both directions.

   Partial Answer: Meridian spacing at 60°S = 3.44 cm

5-2 In Example 5-2, the Bonne projection of Alaska, calculate the remaining chords (for the 50th, 65th, 70th, and 75th parallels).

   Partial Answer: For $\Delta\lambda = 30°$, $\phi = 50°$, chord = 26.70 cm

5-3 Calculate tangent offsets for plotting the 50th parallel in Prob. 5-2. How wide would the sheet of paper have to be if all of this parallel (30° each way) is to be shown? Examine an existing map of Alaska and state whether it would be acceptable to allow the ends of the parallel to fall off the sheet.

   Partial Answer: For $\Delta\lambda = 20°$, y = 2.66 cm

5-4 The height of the sheet of paper for the map of Alaska, Example 5-2, would be 34. 91 cm plus the tangent offset of the 75th parallel. Find this height.

Examine an existing map of Alaska. Would it be reasonable to construct the 73rd parallel instead of the 75th? (The 73rd would be used as an aid in the plotting of meridians, etc. , but then erased.)

Answer: 37. 20 cm

5-5 For a Bonne projection of the continental United States, use 40°N as the central parallel and 95°W as the central meridian. Adopt a scale of 1:10, 500, 000 and calculate the radii needed for plotting the 25th, 30th, 40th, and 50th parallels, and the chords needed for plotting the meridians at 65°W and 125°W. Use centimeters or inches as assigned. Draw the graticule, if assigned, showing meridians at 70°, 80°, 90°, 100°, 110°, and 120° (erasing the ones at 65°, 95°, and 125°).

Partial Answer: For $\phi = 25°$, chord = 28. 66 cm

5-6 Construct a polyconic projection for North America, extending from 25°W to 175°W and from Equator to pole. Begin by drawing the central meridian (100°W) as a vertical line 20 cm in length. Plot points at 15°N, 30°N, 45°N, 60°N, and 75°N. Calculate the various values of R cot ϕ and plot the centers for drawing the parallels. Complete the graticule as described in the text.

Partial Answer: R cot 75° = 3. 41 cm

5-7 Construct a polar orthographic projection of a hemisphere using R = 4 in. , 10 cm, or some other assigned value. Show every 15th parallel and meridian. Construct it graphically or mathematically, as assigned. (Note that the draftsman's plastic triangles may be used in combination to draw all of the meridians.)

REFERENCES

1. C. H. Deetz and O. S. Adams, Elements of Map Projection, Special Publication No. 68, U.S. Coast and Geodetic Survey (now National Geodetic Survey), Washington, D.C. , 1944.

2. J. A. Steers, An Introduction to the Study of Map Projections, 14th Edition, University of London Press, 1965.

3. Erwin Raisz, Principles of Cartography, McGraw-Hill Book Co. , 1962.

4. Tables for the Polyconic Projection of Maps, Special Publication No. 5, U.S. Coast and Geodetic Survey (now National Geodetic Survey), Washington, D.C.

6

CONFORMAL PROJECTIONS
WITH STRAIGHT MERIDIANS

The three projections in Chap. 4 (cylindrical equal-area, Albers, and polar azimuthal equal-area) were modifications of three in Chaps. 2 and 3 (cylindrical equidistant, conical equidistant with two standard parallels, and polar azimuthal equidistant). The modifications were made in such a way that the equal-area property was achieved.

In this chapter, the same three projections will be modified to make them become conformal instead of equal-area. It will be recalled that no projection can be both (Sec. 1-12).

MERCATOR PROJECTION

6-1 CONSTRUCTION

The cylindrical equidistant, with its graticule of perfect squares, is far from being conformal. The scale factor along the meridians is 1.000 whereas along the parallels it is equal to sec ϕ, which is greater than 1.000 at all latitudes except 0°. The Mercator projection allows the scale factor along the meridians to vary with sec ϕ also. At a latitude of 60°, the scale factor is 2.000 in all directions.

The location of a particular parallel may be computed by calculus or taken from published tables.

Assuming the earth to be a sphere, the mathematics may be explained as follows. Figure 6-1 shows dy, the distance between two closely spaced parallels. On a globe of radius R this spacing is

$$dy = R \, d\phi$$

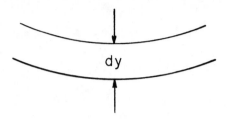

FIG 6-1 An increment of distance between two parallels on a globe.

where dϕ is an increment of latitude. On a Mercator projection this spacing is increased by the local scale factor, sec ϕ.

$$dy = R \sec \phi \, d\phi$$

The distance from the Equator to a particular parallel is a summation of such dy quantities, which, in the language of calculus, involves an integration.

$$y = \int R \sec \phi \, d\phi$$

$$= R \int \sec \phi \, d\phi$$

$$= R \ln (\sec \phi + \tan \phi)$$

$$= R \ln \tan \left(45° + \frac{\phi}{2}\right) \tag{1}$$

Pocket calculators which have trig functions and a natural logarithm key can provide y values very quickly. Table 6-1 was computed this way using R = 1. It also lists the local scale factor (sec ϕ).

For large-scale maps where a spheroid is chosen as the datum, it is convenient to refer to special tables. Reference 1 provides tables for every minute of latitude up to 80° based upon a spheroid with a "flattening" of 1/294. (Flattening is the difference between the equatorial radius and the polar semidiameter divided by the polar semidiameter.) Table 6-2 is extracted from those tables.

Example 6-1

Find the distance from the Equator to the 25th parallel on a Mercator projection if R = 10 cm.

Solution

Assuming the earth to be a sphere, the answer is

$$y = R \ln \tan \left(45° + \frac{\phi}{2}\right)$$

$$= 10 \ln \tan 57.5° = 4.51 \text{ cm}$$

which is less than halfway between the values obtainable for 20° and 30° from Table 6-1.

TABLE 6-1

Mercator Projection Based on Spherical Datum. Distances of the parallels from
the Equator in terms of the radius of the globe

Lat.	Scale factor (sec ϕ)	y	Lat.	Scale factor (sec ϕ)	y
10°	1.015	0.1754R	60°	2.000	1.3170R
20°	1.064	0.3564R	70°	2.924	1.7354R
30°	1.155	0.5493R	80°	5.759	2.4363R
40°	1.305	0.7629R	85°	11.473	3.1313R
50°	1.556	1.0107R	90°	Infinity	Infinity

TABLE 6-2

Mercator Projection Based on Spheroid. Distances of the parallels from the Equator in degrees of longitude at the Equator

0°	0.000	30°	31.278	60°	75.119
5°	4.972	35°	37.182	65°	85.960
10°	9.984	40°	43.461	70°	99.066
15°	15.074	45°	50.224	75°	115.796
20°	20.286	50°	57.610	80°	139.202
25°	25.669	55°	65.814	90°	Infinity

Example: If a degree of longitude at the Equator will equal 4.0 mm at the globe
scale, the values above would simply be multiplied by 4.0 mm.

If the datum is to be the spheroid, Table 6-2 may be used. The length of a degree of longitude along the Equator, if its radius is 10 cm, will be

$$\frac{2\pi R}{360} = 0.1745 \text{ cm}$$

and the required distance will be

y = (25.669) (0.1745) = 4.48 cm

At this scale the difference is 0.3 mm (or about 0.01 inch).

6-2 APPLICATIONS

As shown in Table 6-1, the scale factor becomes very large indeed away from the
Equator. An area located at $\phi = 60°$ will be enlarged by a factor of 2.000 in both
dimensions, thus appearing four times as large as on the globe. At $\phi = 80°$, areas

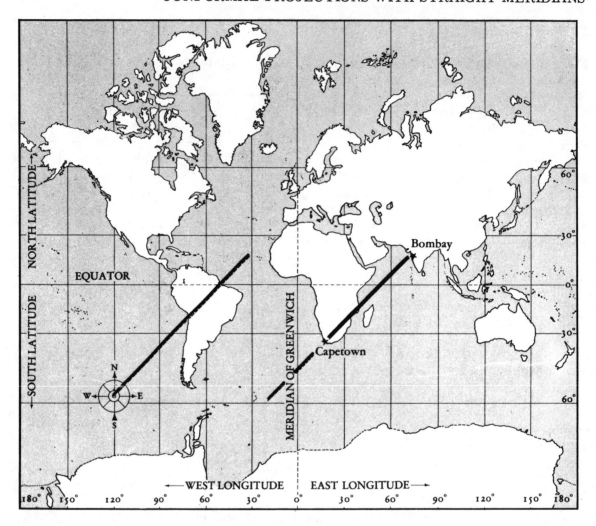

FIG 6-2 Mercator projection showing loxodromes as straight lines and showing great exaggeration of areas away from the Equator. Greenland actually is only one-tenth the size of South America but looks larger here. (From Ref. 2, courtesy of General Dynamics.)

are enlarged by a factor of 33. Despite this extremely poor representation of area, the Mercator projection has great importance and is widely used.

The property of conformality is discussed in Sec. 1-12. Having a single scale factor in all directions, a conformal projection has no angular deformation or distortion of shapes in the vicinity of any point. Thus a Mercator projection will show a given region with correct angular relationships, provided that it doesn't extend very far in the north-south direction into areas with significantly different scale factors. The Times Atlas of the World uses it for a map of Thailand, Burma, and Indo-China. The area exaggeration with respect to the Equator is not important because the Equator does not appear on the map.

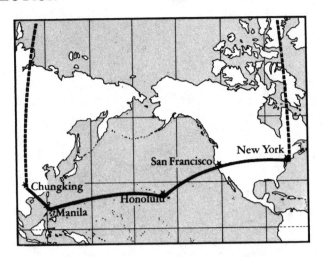

FIG 6-3 Great circle routes plotted on a Mercator projection. (From Ref. 2, courtesy of General Dynamics.)

The Mercator projection has a unique property which has made it useful in navigation for 400 years. A loxodrome or rhumb line, which is a route of constant bearing (Sec. 1-6), is displayed as a straight line. Figure 6-2 shows two loxodromes on a world map. A pilot could use a bearing of N45°E to get from Capetown to Bombay.

Figure 6-3 shows several great circle routes plotted on a Mercator projection. The shortest route from Chungking to New York, which passes almost over the North Pole, cannot be shown in full because the pole would be plotted at infinity. To follow the loxodrome route a pilot would use a constant bearing of N85°E and not go to the pole at all. From Manila to Honolulu the loxodrome would be a sufficiently good approximation of the great circle route while being an easier course to follow. For a longer trip the route can be broken into several constant bearings. (In the case of Chungking to New York, two bearings would be enough: due north to the pole, then due south to New York.)

Chapter 7 discusses a projection which shows great circles as straight lines. That projection (the gnomonic) is often used in conjunction with the Mercator to lay out a route involving several constant bearings which closely approximates the shortest route. Azimuthal projections also show great circles as straight lines, but only if the great circles pass through the center. Figure 6-4 shows the Chungking-New York route on a polar orthographic projection.

The construction of the Mercator was mathematical and did not really involve projecting to a cylinder. However, a Mercator map could be rolled around a globe and become a cylinder tangent along the Equator. Such a cylinder would be co-axial with the globe. Chapter 10 discusses an important variation, the transverse Mercator, in which the "cylinder" is turned 90° to the axis of the globe.

Several projections have been devised which resemble the Mercator but which show less exaggeration of area. Among them are Gall's [3] and Miller's [4].

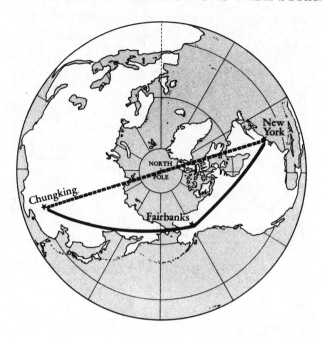

FIG 6-4 Great circle routes plotted on a polar orthographic projection.   (From Ref. 2, courtesy of General Dynamics.

LAMBERT CONFORMAL CONIC PROJECTION

On the conical equidistant with two standard parallels, the meridians were standard lines.   The Albers projection was described as a variation in which the scale factor along each meridian varied from 1. 000 so as to be compensatory with respect to the east-west scale factor.   The Lambert conformal conic is a variation in which the scale factor along each meridian varies so as to be equal to the east-west scale factor.   This condition makes the standard parallels slightly closer together than in the other two cases (see Fig. 6-5).

6-3 CONSTRUCTION

In Sec. 1-7 and in the discussions of the various conic projections the "constant of the cone" k was introduced.   It is a decimal number (less than 1. 00) expressing how much of a full circle is required to display 360° of longitude, or how large the central angle L is in comparison to $\Delta\lambda$, the longitudinal extent of the map. If the longitudinal extent is 360° (as in a map of a hemisphere), k = L/360°.

The expression for k for the Lambert conformal conic is given here without proof [5].   Assuming the datum to be a sphere,

$$k = \frac{\log \cos \phi_s - \log \cos \phi_n}{\log \tan \dfrac{90 - \phi_s}{2} - \log \tan \dfrac{90 - \phi_n}{2}} \qquad (2)$$

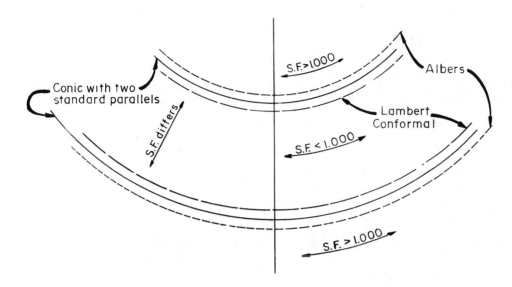

FIG 6-5 Comparison of three conic projections. The distance between the standard parallels depends upon whether the north-south scale factor is standard (equidistant), or either compensatory to or equal to the east-west scale factor.

in which ϕ_n is the greater of the two latitudes chosen to be standard. Another constant of the projection, called c, is needed. Again without proof, the expression is

$$c = \frac{R \cos \phi_n}{k\left(\tan \frac{90 - \phi_n}{2}\right)^k} \quad or \quad \frac{R \cos \phi_s}{k\left(\tan \frac{90 - \phi_s}{2}\right)^k} \tag{3}$$

The radius for plotting any parallel is

$$r = c\left(\tan \frac{90 - \phi}{2}\right)^k \tag{4}$$

where ϕ is the latitude of the parallel in question. The parallels are concentric
circles and the meridians are radial straight lines with central angles derived
from k.

Example 6-2

Plan a graticule for Scandinavia similar to the one discussed in Example 4-2,
but using the Lambert conformal conic instead of the Albers projection. As
before, R = 100 cm, ϕ_n = 68°, ϕ_s = 58°, and the region is bounded by 4°E,
32°E, 54°N, and 72°N.

Solution

The first step, as with the Albers, is to find the constant of the cone, k.

$$k = \frac{\log \cos 58° - \log \cos 68°}{\log \tan 16° - \log \tan 11°} = 0.89215$$

which is slightly larger than for the Albers (0.88762).

The other constant c is found next.

$$c = \frac{100 \cos 68°}{0.89215 \ (\tan 11°)^{0.89215}} = 181.04$$

which may be found again using ϕ_S as a check.

The graticule may be plotted by coordinates (tangent offsets), but because the region is so far north, it may be convenient to use a beam compass and scale. The radii are as follows:

$$r_{54} = 181.04 \ (\tan 18°)^{0.89215} = 66.41 \text{ cm}$$

$$r_{56} = 181.04 \ (\tan 17°)^{0.89215} = 62.90 \text{ cm}$$

$$r_{58} = 181.04 \ (\tan 16°)^{0.89215} = 59.40 \text{ cm}$$

etc.

To verify that the standard parallels are indeed closer together than in the Albers, as shown in Fig. 6-5, one may calculate $r_{58} - r_{68}$, which is 59.40 - 41.99 = 17.41 cm, and compare it with $y_{68} - y_{58}$ in the first column of Table 4-1. The latter is 24.46 - 6.96 = 17.50 cm. If the meridians were standard, the distance would be $(10/180) \pi R = 17.45$ cm.

To plot the limiting meridians, which are 14° on each side of the central meridian, a chord distance may be calculated for one of the parallels. Using the 54th parallel,

$$\text{Chord} = 2(66.41) \sin \frac{14°k}{2} = 14.45 \text{ cm}$$

This parallel may be divided into equal parts by trial, using dividers. Radial lines may then be drawn to each point to form the meridians.

6-4 APPLICATIONS

The Lambert conformal conic is the most important of the several projections developed by J. H. Lambert. It is widely used in atlases, for aeronautical charts, and for plane coordinate systems in surveying. Although developed in 1772, it was rarely used until World War I, when the French army adopted it for battle maps. It offered a good combination of minimum angular and scale distortion.

Because of the difficulty of the computations, especially on the spheroid, tables have been published for several combinations of standard parallels [6] and computer programs are available [7]. The equations given here, based on the sphere, are easily handled on pocket calculators.

The value of this and other conformal projections as a basis for plane coordinate systems in surveying work results from two related features:

1. Distance measurements in a survey can be multiplied by a single scale factor regardless of the direction of the line.

2. Angle measurements in a survey require no corrections at all (except for unusually long lines in highly precise surveys).

POLAR STEREOGRAPHIC PROJECTION

Like the other polar azimuthal projections covered earlier, the stereographic shows meridians radiating from the pole at equal angles. Meridians 15° apart on the globe are 15° apart on the map. The spacing of the parallels, instead of being equidistant (Chap. 2), equal-area (Chap. 4), or orthographic (Chap. 5), is designed to make the projection be conformal.

6-5 CONSTRUCTION

The stereographic is literally projected, like the cylindrical equal-area and orthographic projections, rather than being mathematically created. Figure 6-6 shows that a tangent plane is used and that the rays of projection emanate from a point diametrically opposite to the point of tangency. In the polar case, the rays extend from the opposite pole. From similar triangles it can be seen that the Equator will be a circle exactly twice as large as on the globe.

The radii for drawing the various parallels may be obtained graphically or calculated from the equation shown in Fig. 6-6.

Example 6-3

Verify that a stereographic projection is conformal by calculating an approximate scale factor in two perpendicular directions at $\phi = 60°$.

Solution

The scale factor along the 60th parallel may be determined by dividing its length on the map by its length on the globe.

$$\text{Map distance} = 2\pi\left(2R \tan \frac{90 - \phi}{2}\right)$$

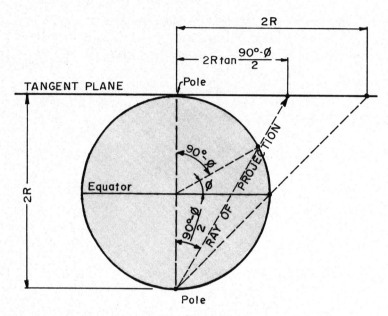

FIG 6-6 The stereographic projection is projected to a tangent plane from the diametrically opposite point on the globe.

Globe distance $= 2\pi R \cos\phi$

Scale factor $= \dfrac{2\tan\left[(90-\phi)/2\right]}{\cos\phi}$

$= \dfrac{2\tan 15^\circ}{\cos 60^\circ} = 1.0718$

In the radial direction, consider the 1° increment from $\phi = 59.5^\circ$ to $\phi = 60.5^\circ$, and divide map distance by globe distance.

Map distance $= 2R\tan\left(\dfrac{90-59.5}{2}\right) - 2R\tan\left(\dfrac{90-60.5}{2}\right)$

$= 2R\,(\tan 15.25^\circ - \tan 14.75^\circ)$

Globe distance $= \dfrac{\pi R}{180}$

Scale factor $= \dfrac{2(\tan 15.25^\circ - \tan 14.75^\circ)\,180}{\pi}$

$= 1.0718$

6-6 APPLICATIONS

The military grid system is discussed in Chap. 10. In the polar regions it utilizes the polar stereographic projection in a system known as UPS, Universal Polar Stereographic. Except for this application, where conformality is important, the other polar azimuthal projections are more useful.

The case of a stereographic projection centered on the Equator and the oblique case are covered in Chap. 8. It is interesting to note that the stereographic projection dates back to ancient Greece.

PROBLEMS

6-1 Plan a Mercator projection for a navigational chart of Lake Superior. The map is to be bounded by 46°N, 49°N, 83°W, and 93°W, and the width of the sheet is limited to 52 cm. Assume the earth to be a sphere. Show the dimensions on a sketch.

Partial Answer: Height = 23.10 cm

6-2 For the Lambert conformal conic projection of Example 4-2, calculate the remaining radii (r_{60}, r_{62}, etc.).

Partial Answer: r_{66} = 45.48 cm

6-3 Plan a graticule for a map of Hudson Bay, in Canada, using the Lambert conformal conic projection with standard parallels at 53° and 63°. The scale is 1:8,000,000 and the area is bounded by 50°N, 66°N, 75°W, and 95°W. Plan to show meridians and parallels at 5° intervals (not showing parallels at 53°, 63°, or 66°). Make a sketch or plot the graticule, as assigned. (Problem 2-8 involved a similar map using a different projection.) Note: If your solution is to be used in Prob. 6-4, it is necessary to assume that the radius of the datum is exactly 6370.00 km and to compute k, c, and r_{60} to six significant figures.

Partial Answer: r_{60} = 46.7328 cm

6-4 Assume that a plane (rectangular) coordinate system has been established for the Hudson Bay area using the data in Prob. 6-3. A field measurement across a harbor, made electronically, was 3,030.00 m. If the latitude was 60°, multiply the measurement by an appropriate scale factor to get a "grid distance." This will involve using r_{60} and R from Prob. 6-3.

Answer: 3020.09 m

6-5 The following table lists the radii for drawing the parallels on a polar stereographic projection in which R = 15 cm. Complete the table and comment on the result.

ϕ	r
North Pole	0.00
60°N	8.04
30°N	17.32
Equator	
30°S	
60°S	
South Pole	

Partial Answer: for 60°S, 111.96 cm

6-6 By what factor would <u>areas</u> be exaggerated in the vicinity of 60°N in Prob. 6-5? (See Example 6-3.)

    <u>Answer:</u> 1.15 times

6-7 Discuss the suitability of each of the following projections for a map of Africa showing the distribution of population.

    a. Mercator
    b. Bonne
    c. Sinusoidal

REFERENCES

1. C. H. Deetz and O. S. Adams, <u>Elements of Map Projection</u>, Special Publication No. 68, U.S. Coast and Geodetic Survey (now National Geodetic Survey), Washington, D.C., 1944.

2. <u>Maps and How to Understand Them</u>, Consolidated Vultee Aircraft Corporation, 1943.

3. Erwin Raisz, <u>Principles of Cartography</u>, McGraw-Hill Book Co., New York, 1962.

4. A. H. Robinson and R. D. Sale, <u>Elements of Cartography</u>, 3rd Edition, John Wiley and Sons, 1969.

5. J. A. Steers, <u>An Introduction to the Study of Map Projections</u>, 14th Edition, University of London Press, 1965.

6. Special Publications Nos. 52, 68 and the State Plane Coordinate Projection Tables for many of the states, U.S. Coast and Geodetic Survey (now the National Geodetic Survey), Washington, D.C.

7. Custom Designed Lambert Projection, H794, listed in <u>Annual Report on Cartographic Research</u>, U.S. Geological Survey, June 1976.

GNOMONIC PROJECTIONS

Gnomonic projections are truly projected to a cylinder, cone, or plane from the center of the globe.

The cylindrical case, sometimes called the perspective projection is shown in Fig. 7-1. The extreme exaggeration of areas away from the Equator is evident in the case of Alaska, for example. Because of this distortion and a lack of any special advantages, the cylindrical case is rarely used. It will not be discussed further here.

The conic case also is of no great value in comparison to the other conic projections that are available. Each of the six conics covered in earlier chapters has advantages over a gnomonic.

The gnomonic projection to a plane is a different story. It is very important and will be the subject of the remainder of this chapter. In fact, the name "gnomonic," used alone, is understood to refer to this case. It offers the advantage that all great circles appear as straight lines. This feature is discussed in Sec. 7-4.

THE GNOMONIC

7-1 CONSTRUCTION OF THE POLAR GNOMONIC

If the plane of projection is tangent at a pole, the radius for drawing any parallel may be found graphically or mathematically, as shown in Fig. 7-2. The general appearance of the polar gnomonic is similar to that of the other polar azimuthal projections, but the radii are greater. Figures 7-3 and 7-4 show the comparison.

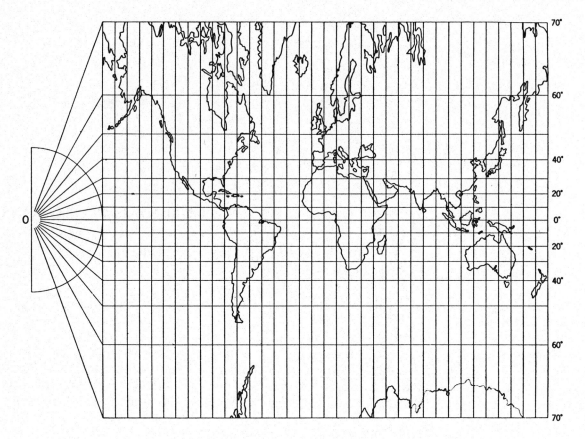

FIG 7-1 Gnomonic cylindrical projection. (From Ref. 1.)

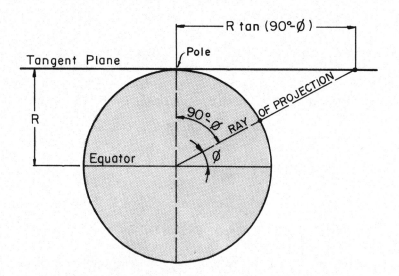

FIG 7-2 Polar gnomonic projection. Note that the Equator would appear at infinity; the map is limited to less than a hemisphere.

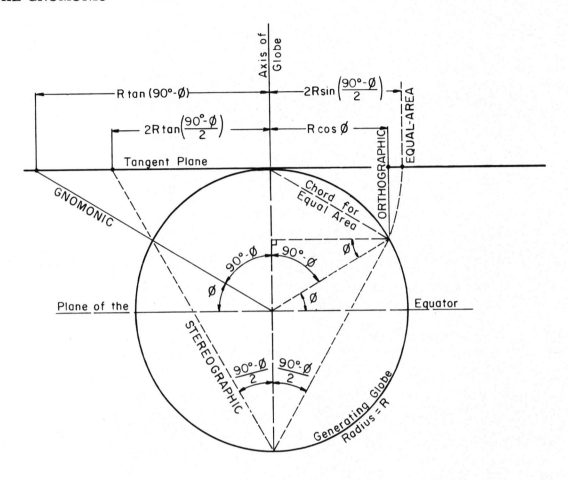

FIG 7-3 Comparison of four polar azimuthal projections. The azimuthal equidistant is not shown.

7-2 CONSTRUCTION OF THE EQUATORIAL CASE

If the plane is tangent at the Equator instead of the pole, the graticule will be projected as shown in Fig. 7-5.

The meridians, being great circles, will all be straight lines, as will the Equator. The points at which the meridians cross the Equator and those at which the parallels cross the central meridian may be plotted using a compass. For example, meridians lying 15° east and west of the central meridian and parallels at 15°N and 15°S may be located by drawing a circle of radius R tan 15° as shown in Fig. 7-6. This is the circle corresponding to the 75th parallel in the polar gnomonic.

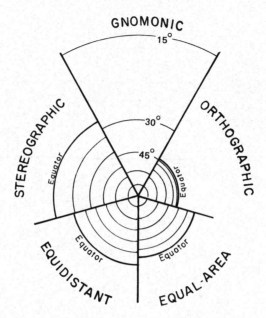

FIG 7-4 Five polar azimuthal projections, all shown at the same scale. Parallels are spaced at 15°. Outermost parallel is the Equator except in the gnomonic, where it is the 15th parallel.

The y coordinate for plotting any parallel ϕ as it crosses any meridian is

$$y = R \sec \Delta\lambda \ \tan \phi \quad \text{or} \quad \frac{R \tan \phi}{\cos \Delta\lambda} \tag{1}$$

where $\Delta\lambda$ is the difference in longitude between the central meridian and the one on which y is to be plotted. This expression is similar to the one used for setting the compass in Fig. 7-6 in that it involves tan ϕ, but the "adjacent side" is R sec $\Delta\lambda$ instead of just R (see Fig. 7-7).

Example 7-1

Plot a gnomonic projection for the Indian Ocean centered at $\phi = 0°$, $\lambda = 80°$E. Use R = 40 cm.

Solution

A plotting table of y values on various meridians may be set up as follows. In Table 7-1 the expression for y is shown with a cosine function instead of secant for convenience in using a pocket calculator. As each row is filled in, the numerator is a constant. Entries in the first column ($\Delta\lambda = 0°$) may be used for plotting the circles shown in Fig. 7-6 to locate the meridians.

The parallels are drawn with the aid of a flexible curve. Intermediate points may be computed for additional precision.

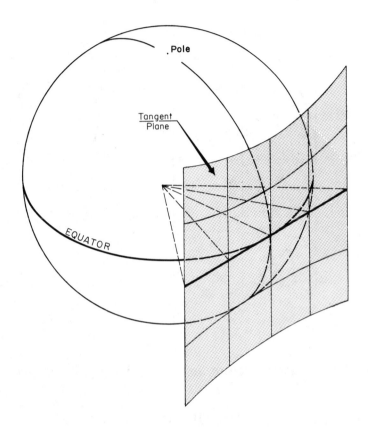

FIG 7-5 The equatorial gnomonic projection and its generating globe.

TABLE 7-1

Equatorial Gnomonic Projection. Data for plotting each quadrant in Example 7-1.

λ =	80°E	65° & 95°	50° & 110°	35° & 125°
Δλ =	0°	15°	30°	45°
ϕ		$y = \dfrac{R \tan \phi}{\cos \Delta\lambda}$		
0°	0.00	0.00	0.00	0.00
15°	10.72	11.10	12.38	15.16
30°	23.09	23.91	26.67	32.66
45°	40.00	41.41	46.19	56.57

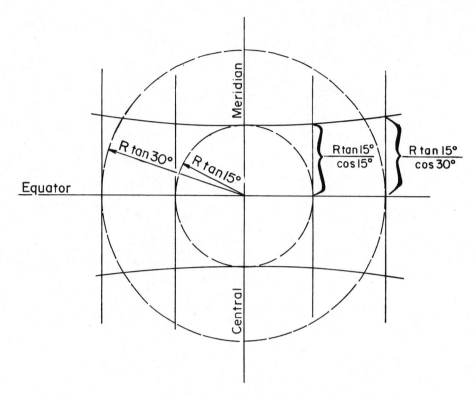

FIG 7-6 Construction of an equatorial gnomonic projection.

7-3 CONSTRUCTION OF THE OBLIQUE CASE

There are several ways to construct a gnomonic projection which has its plane tangent at some point between the Equator and the pole. Reference 2 presents two methods. The method shown here is based on the graphic approach shown in that reference but is a semigraphic variation. By calculating some of the key dimensions, a more precise graticule will be achieved.

To center the projection at $\phi = 58°$N (for Juneau, Alaska), calculate the quantities shown in Fig. 7-8. The map itself, or the graticule, is shown at the top. The Equator, being a great circle, is a straight line. The central meridian, of unknown length, is a perpendicular at Q. The other meridians, also great circles, will be straight lines such as P'K. To locate P' and several points along the Equator like K, construct a generating globe with its axis horizontal as shown. It will show the projection of the central meridian from pole to Equator as Z to Q, tangent to the globe at a latitude of $\phi = 58°$. As can be seen in the figure, the center of the globe 0 must be plotted at a distance of R sec ϕ below Q. Point Z will be important and must be plotted to the left of 0 at a distance of R csc ϕ. Line ZQ should be drawn but need not be scaled except as a check. From the right triangles shown, its length should be R cot ϕ + R tan ϕ. This same distance will locate P'.

Points like K, where the meridians meet the Equator, may be plotted at distances from Q equal to R sec ϕ tan $\Delta\lambda$, where $\Delta\lambda$ is the longitudinal "distance" to

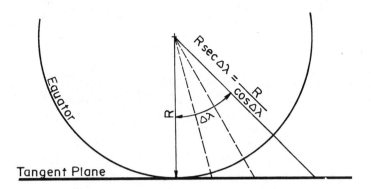

FIG 7-7 The plane of the Equator, showing that the distance to the tangent plane is R sec Δλ.

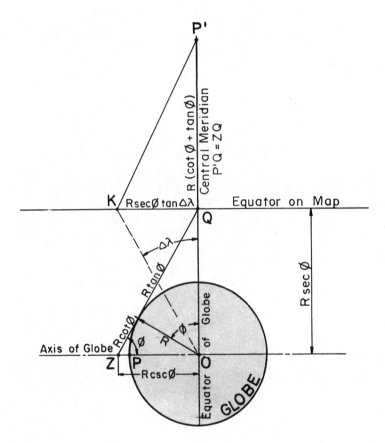

FIG 7-8 Construction of the oblique gnomonic projection.

the meridian from the center. KQ is part of right triangle KQO. For this construction, the circle around O is temporarily viewed as the Equator instead of a meridian.

With the quantities shown in Fig. 7-8 calculated in advance and then plotted, a graphic construction may be used to locate the parallels. This is shown in Fig. 7-9.

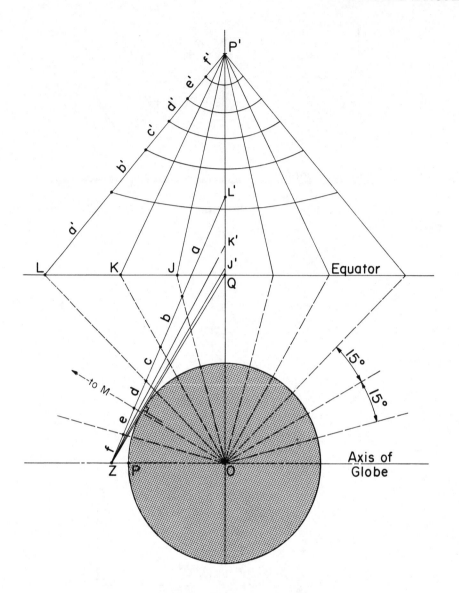

FIG 7-9 Graphic method of locating parallels on oblique gnomonic projection.

To locate the parallels along meridian P'L, first lay out a distance equal to P'L from Z in the lower part of Fig. 7-9. This is ZL'. Radial lines such as OJ, OK, OL, and OM will cut line ZL' at intervals shown as a, b, c, d, etc. These may be transferred by dividers to the graticule where they are shown as a', b', c', d', etc.

Other meridians are divided in a similar manner. A flexible curve may be used to draw the parallels.

Example 7-2

Plot an oblique gnomonic projection centered on Juneau, Alaska, which is at $\phi = 58°N$, $\lambda = 135°W$. Use R = 10 cm.

Solution

First solve for the quantities shown in Fig. 7-8 as follows:

$$OQ = R \sec \phi = \frac{R}{\cos \phi} = \frac{10}{\cos 58°} = 18.87 \text{ cm}$$

$$OZ = R \csc \phi = \frac{R}{\sin \phi} = \frac{10}{\sin 58°} = 11.79 \text{ cm}$$

$$ZQ = P'Q = R(\cot \phi + \tan \phi) = 10\left(\frac{1}{\tan 58°} + \tan 58°\right) = 22.25 \text{ cm}$$

Also, of course, $ZQ = (OZ^2 + OQ^2)^{1/2}$.

Finally, the distances from Q to J, K, L, and M are equal to OQ $(\tan \Delta\lambda)$ for each value of $\Delta\lambda$.

For 15°:   QJ =   5.06 cm

For 30°:   QK = 10.90 cm

For 45°:   QL = 18.87 cm

For 60°:   QM = 32.69 cm

The graphic procedure is as shown in Fig. 7-9. The map is symmetrical about the central meridian.

7-4 APPLICATIONS

The unique property of the gnomonic projection (meaning the azimuthal projection, not the cylindrical or conic ones) is that it displays all great circles as straight lines. This is mentioned in Sec. 6-3, where the Mercator is discussed. The fact that the projection has this property may be understood by remembering that great circles lie on planes which include the center of the globe and that the center of the globe also is the point from which the gnomonic is projected. A similar situation occurs in engineering astronomy, where the stars in the sky are considered to lie on an imaginary "celestial sphere." The surveyor views the sky from the tiny earth at the center and sees all great circles as straight lines. He is in the plane of each one (the celestial meridians and celestial equator, for example.

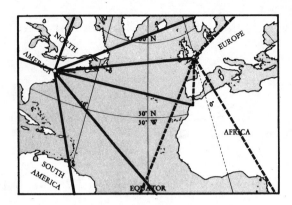

FIG 7-10 Gnomonic chart of the North Atlantic showing great circles.   (From Ref. 3, courtesy of General Dynamics.)

Figure 7-10 shows eleven great circles plotted on a gnomonic chart of the North Atlantic. Figure 7-11 shows seven of them transferred onto a Mercator projection, where they are no longer straight. An air navigator could plan a flight from Washington to Moscow on the Mercator by laying out three loxodromes, the first one going to the tip of Greenland, the second one going to the south coast of Iceland, and the third one going to Moscow. He would have three constant bearings to follow which would give him a flight approximating a great circle.

Figures 7-3 and 7-4 show that the gnomonic is always limited to less than a hemisphere. The U.S. Navy and the British Admiralty have issued gnomonic charts for the various oceans on separate sheets, namely, one each for the North Atlantic, South Atlantic, Pacific, North Pacific, South Pacific, and Indian Oceans. The projection also has been used for smaller regions such as harbors, but for such limited areas it differs little from the Mercator.

It should be remembered that any of the azimuthal projections, if centered on Washington, D.C., would show all of the great circles of Fig. 7-11 as straight lines. The azimuthal equidistant, in fact, will also show their correct lengths. This is shown in Fig. 7-12. The azimuthal equidistant will be covered in Chap. 8.

PROBLEMS

7-1 Find the scale factor along the 60th parallel of a polar gnomonic projection. Compare it to the value found for the polar stereographic in Example 6-3, namely, 1.0718.

   Partial Answer: Scale factor is greater than in Example 6-3

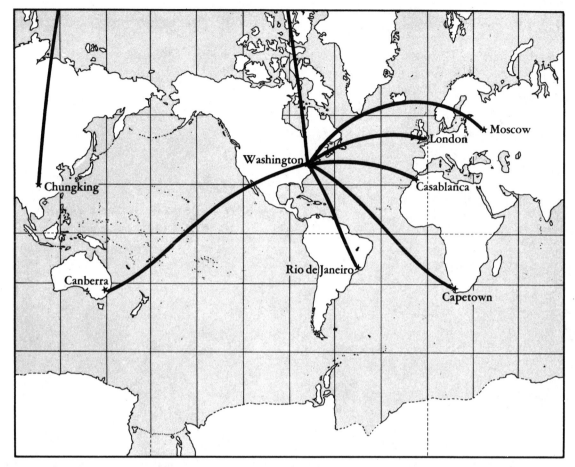

FIG 7-11 Mercator projection showing great circle routes from Washington, D.C. (From Ref. 3, courtesy of General Dynamics.)

7-2 Extend Table 7-1 to include $\phi = 60°$ and $\Delta\lambda = 60°$. Note the increasing intervals in each column.

    <u>Partial Answer:</u> For $\phi = 45°$, $\Delta\lambda = 60°$, y = 80.00

7-3 Draw Fig. 7-7 to scale using R = 10 cm. Add a line perpendicular to the line labeled R sec $\Delta\lambda$ at the point where it meets the tangent plane. Make a right triangle by adding another radial line with a central angle of 15°. Scale the short leg of the triangle to verify that it is equal to $y = R \sec 45° \tan 15°$. Label it as y_{15}. (This illustrates a graphic method of constructing the equatorial gnomonic.)

7-4 Complete the graphic part of Example 7-2 to construct the oblique gnomonic projection for Juneau, Alaska. Use the data shown in the solution to determine the size of paper required. Note that it is always possible, in graphic solutions, to attach extra paper temporarily or to complete the construction on a large sheet of scrap paper and then to trace only the graticule on a smaller sheet.

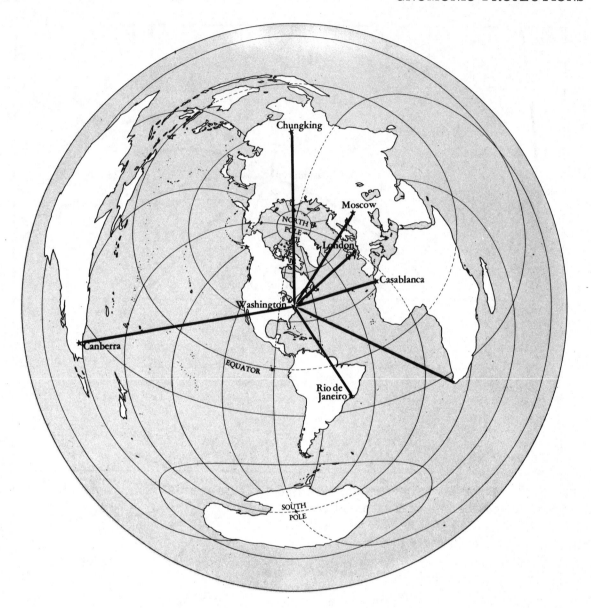

FIG 7-12 Azimuthal equidistant projection centered on Washington, D.C., showing same great circles as Figs. 7-10 and 7-11. (From Ref. 3, courtesy of General Dynamics.)

7-5 On the graphic construction of Prob. 7-4, carefully add Salt Lake City for
     which ϕ = 40.5°N and λ = 111.5°W. Do this by calculating a distance simi-
     lar to QK for the meridian through Salt Lake City. Call it QS. On the lower
     half of your construction, lay out the distance equal to P'S and obtain the dis-
     tance needed to plot the city on its meridian.

     <u>Partial Answer:</u> QS = 8.20 cm

7-6 Construct an oblique gnomonic projection centered on London for which $\phi = 51°N$ and $\lambda = 0°$. Use $R = 12$ cm and plot the graticule at $10°$ intervals out to $40°$ east and west of London and from $10°N$ to the pole.

   Partial Answer: QJ = 3.36 cm

7-7 It is possible to fit a cube around a globe, tangent at six points. Six gnomonic projections can be plotted on the faces of the cube. If four of the projections are equatorial, two will be polar. For this arrangement, plot one of the equatorial projections for a cube measuring 30 cm on a side. (It will cover one-quarter of the Equator, or $90°$ of longitude.)

REFERENCES

1. C. H. Deetz and O. S. Adams, Elements of Map Projection, Special Publication No. 68, U.S. Coast and Geodetic Survey (now National Geodetic Survey), Washington, D.C., 1944.

2. J. A. Steers, An Introduction to the Study of Map Projections, 14th Edition, University of London Press, 1965.

3. Maps and How to Understand Them, Consolidated Vultee Aircraft Corporation, 1943.

8

ADDITIONAL AZIMUTHAL CASES

All five of the common azimuthal projections are introduced in previous chapters, but usually just the polar case is presented. Only for the gnomonic are equatorial and oblique cases explained. This chapter discusses the equatorial and oblique cases of the stereographic and a method by which they can be used to form any other azimuthal projection. Also there will be a brief discussion of the oblique cases of the azimuthal equidistant and azimuthal equal-area projections including modern ways of constructing them.

EQUATORIAL STEREOGRAPHIC PROJECTION

8-1 CONSTRUCTION

All cases of the stereographic projection can be constructed with ease and precision because all parallels and meridians are either circular or straight. There are no irregular curves. The equatorial case will be explained for one hemisphere, that being its most common application.

The bounding meridian is a circle of radius r = 2R, like that used for the Equator in the polar case. Within this circle, the central meridian and Equator are drawn as straight lines, as in Fig. 8-1. The other meridians and parallels are circles whose centers may be found graphically or mathematically.

When the bounding meridian is projected on the tangent plane, the rays of projection come from a point on the Equator that is diametrically opposite to the center of the map (see Fig. 6-6). While the bounding meridian in Fig. 8-1 is thus doubled in size, a point Q that was 60° above the Equator will continue to be at that angle on the map. In other words, the parallels will meet this bounding circle at equal intervals, such as 10° apart, just as on the original meridian. Because

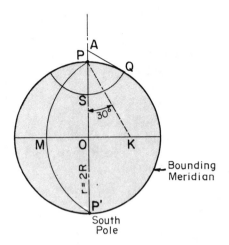

FIG 8-1 Construction of the equatorial stereographic projection of a hemisphere.

the projection is conformal, the parallels must intersect the bounding meridian (and all others) at right angles. The center for drawing the 60th parallel is at A, found by constructing AQ tangent to the circle (or perpendicular to radius OQ). To find A mathematically, distance OA may be calculated as follows:

$$OA = OS + AQ$$

$$= 2R \tan \frac{\phi}{2} + r \tan (90 - \phi) \tag{1}$$

The first term may be seen by drawing a "side view" of the globe showing OS on a tangent plane and the ray of projection coming from the point opposite from O. The second term comes from Fig. 8-1 in the triangle formed by A, Q, and O (line OQ is not drawn in). The expression could be simplified slightly if desired by substituting r for 2R.

To draw a meridian lying 60° west of the central meridian, another center K must be found. The desired meridian PMP' meets the central meridian at a 60° angle because angles at all points are preserved in conformal projections. Thus angle KPO = 90° - 60° or 30°, and distance OK is r tan 30°. It is equal to AQ. The meridian is a circle drawn through the poles using K as the center. Distance MO, though not needed, is equal to OS.

At the time the compass is set for distance AQ, it is used also to mark point K and to draw the other 60th parallel (in the Southern Hemisphere). It should also be used to plot a point comparable to K on the left side of the central meridian.

Example 8-1

Plan an equatorial stereographic projection of a hemisphere to fit within a 40-cm square. Show meridians and parallels at 30° intervals.

Solution

The radius for drawing the bounding meridian is r = 20 cm. The radius of the generating globe R = 10 cm, and the scale is

$$\text{RF scale} = \frac{10}{637,000,000} = \frac{1}{63,700,000}$$

Points corresponding to centers A and K and point S are found for parallels and meridians at 30° and 60°.

ϕ	$OS = 2R \tan \frac{\phi}{2}$	$OK = AQ = r \tan (90 - \phi)$	$OA = OS + AQ$
0°	0.00	Infinity	Infinity
30°	5.36	34.64	40.00
60°	11.55	11.55	23.09
90°	20.00	0.00	20.00

Another way of plotting this graticule is with the following mapping equations from Ref. 1. They may be solved by programmable calculator or computer.

$$x = \frac{r \cos \phi \sin \lambda}{1 + \cos \phi \cos \lambda} \tag{2}$$

$$y = \frac{r \sin \phi}{1 + \cos \phi \cos \lambda} \tag{3}$$

The same reference gives the equations of the circular meridians and parallels, and an equation for the scale factor. For distance OA in Fig. 8-1 it gives r csc ϕ instead of the expression used here. It also gives a method for basing the projection on the spheroid rather than the sphere.

8-2 APPLICATIONS

The equatorial stereographic is often used in atlases to show hemispheres (see Fig. 8-2). It is easy to construct and conformal. Another name in use is the "stereographic meridian projection."

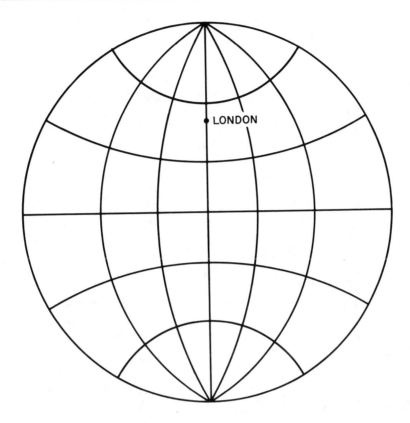

FIG 8-2 Equatorial stereographic projection with 30° intervals. (Drawn by
L. Stephen Hawbaker.)

OBLIQUE STEREOGRAPHIC PROJECTION

8-3 CONSTRUCTION

As stated in Sec. 8-1, the oblique stereographic projection also consists only of
circles and straight lines. It is easier to construct than other oblique cases.

The principles of this projection will be presented but a few equations will be
given without proof. Figure 8-3 shows a generating globe at left and a plane that
is tangent at 45°N. (This value is called ϕ_O, the latitude of the origin.) The axis
of the globe and the plane of the Equator are shown, as well as V, the point opposite to the origin, from which the rays of projection emanate. Five points have
been projected to the plane and have been transferred to the central meridian of
the map, shown at the right. These points are the two poles, the two crossings
of the Equator, and a point that is as far below the Equator as the origin is above
it. The latter point is on the homolatitude (or "same latitude"), which is a parallel that will project as a straight line because it contains the point V.

Because the projection is conformal, all meridians must cross the homolatitude at right angles. Thus their centers must lie on that line. Also, of course,
the meridians must pass through both poles. One meridian is shown, drawn
around its center K. All of the meridians will meet the central meridian at the

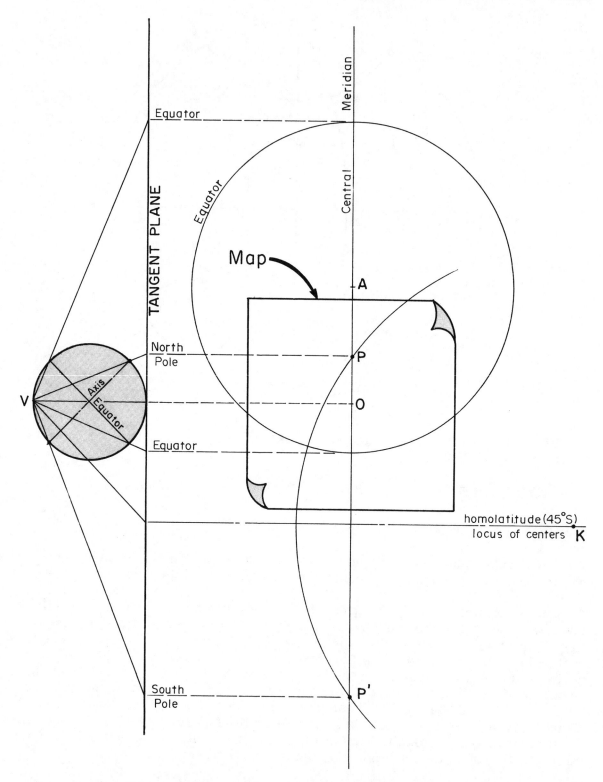

FIG 8-3 Principles of the oblique stereographic projection.

same angle as they do on the globe; thus the meridians that lie 90° east and west of the central meridian will be perpendicular to the central meridian, or horizontal on the map sheet, at both poles. Point K, in this case, will be on the central meridian, midway between the poles. In general, K may be found by knowing the direction of the meridian at P and drawing a radial line perpendicular to it from P toward K. The intersections with the homolatitude may be obtained graphically or mathematically.

The Equator, and all other parallels, must cross the central meridian at right angles. Their centers must lie on that meridian midway between the two crossings. The Equator has been drawn around A as the center. The map sheet has been shown as a square of arbitrary size. Figure 8-4 shows a completed graticule bounded by a circle instead. It is one hemisphere, centered on London (51°N).

From the above relationships it can be shown [1] that the meridians may be plotted with the following equations: Location of centers (K) will be at $x = -r \sec \phi_0 \cot \lambda$ and $y = -r \tan \phi_0$, and the radii will be $r_\lambda = r \sec \phi_0 \csc \lambda$. Similarly, the parallels may be drawn as follows: Location of centers (A) will be at $x = 0$ and $y = r \cos \phi_0/(\sin \phi_0 + \sin \phi)$, and the radii will be $r_\phi = r \cos \phi/(\sin \phi_0 + \sin \phi)$. Some of the construction will extend far beyond the map sheet, especially for meridians close to the central one. To avoid this problem, some or all of the graticule intersections may be plotted from the following mapping equations taken from Ref. 1.

$$x = \frac{r \sin \lambda \cos \phi}{1 + \sin \phi \sin \phi_0 + \cos \phi \cos \phi_0 \cos \lambda} \tag{4}$$

$$y = \frac{r (\sin_0 \cos \phi_0 - \sin \phi_0 \cos \phi \cos \lambda)}{1 + \sin \phi \sin \phi_0 + \cos \phi \cos \phi_0 \cos \lambda} \tag{5}$$

A program for solving these last two equations on an HP-25 pocket calculator is given in Appendix A-2. It takes advantage of the fact that the denominator is the same in both equations. The distance r is the same as it was in Sec. 8-2, namely, the radius for drawing a hemisphere, 2R. Another program, given in Appendix A-4, handles similar equations using an SR-56 (Texas Instruments).

Example 8-2

Verify that the equations given for drawing parallels will indeed result in the circle shown for the Equator in Fig. 8-3. For convenience, use R = 1.0 unit.

Solution

First, use Fig. 8-3 instead of the equations. The left half of the figure shows that the Equator is 45° below the point of tangency and also 135° above it.

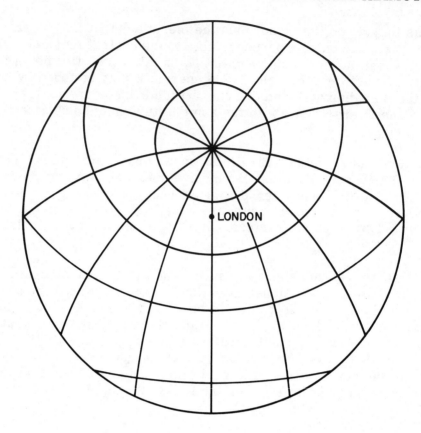

FIG 8-4 Stereographic projection of a hemisphere, centered on London. (Drawn by L. Stephen Hawbaker.)

These two points project to the central meridian at the following y coordinates, according to the geometry of the figure:

$$y_{lower} = -2R \tan 22.5° = -0.828R = -0.828 \text{ units}$$

$$y_{upper} = +2R \tan 67.5° = +4.828R = +4.828 \text{ units}$$

This geometry also may be seen in Fig. 6-6. The radius for drawing the Equator is half of the distance between these points.

$$r_\phi = \frac{4.828 + 0.828}{2} = 2.828 \text{ units}$$

The y coordinate of A is

$$y_A = 2.828 - 0.828 = 2.000 \text{ units}$$

Now using the given equations for these same quantities,

$$r_\phi = \frac{r \cos \phi}{\sin \phi_0 + \sin \phi} = \frac{2R \cos 0°}{\sin 45° + \sin 0°} = 2.828 \text{ units}$$

$$y_A = \frac{r \cos \phi_0}{\sin \phi_0 + \sin \phi} = \frac{2R \cos 45°}{\sin 45° + \sin 0°} = 2.000 \text{ units}$$

8-4 APPLICATIONS

The oblique stereographic projection is used for plane coordinate systems in some jurisdictions including two provinces of Canada (New Brunswick and Prince Edward Island). When used for a hemisphere, as in Fig. 8-4, it provides a conformal projection bounded by a circle of uniform scale factor (namely 2.0). This combination of properties makes it the best of the conformal projections for a hemisphere [2]. Another name for the oblique case is the "stereographic horizon projection."

The following section will present a way to use this projection to construct other oblique azimuthal projections.

TRANSFORMING AZIMUTHAL PROJECTIONS

8-5 PRINCIPLE

Chapter 7 offers a comparison of the five most common azimuthal projections in their polar cases (Figs. 7-3 and 7-4). The only difference was in the radius used in drawing the parallels.

The similarity exists in the equatorial and oblique cases as well. If various azimuthal projections are centered on London, they will all show the great circle route to Rome as a straight line in the correct direction (hence the name "azimuthal"), but the distance will vary. It will be correct on the azimuthal equidistant, too short on the orthographic, too far on the gnomonic, and so on. If the central angle (arc distance) is obtained, using the equation for cos D in Sec. 1-6, the map distance to Rome may be calculated for each different projection using the equations shown in Fig. 7-3. In that figure, for the polar cases, the central angle was 90 - ϕ.

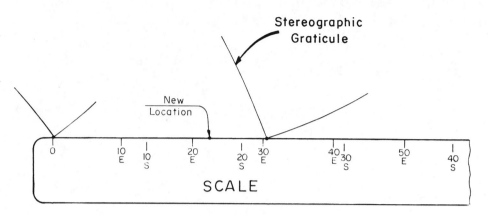

FIG 8-5 Scale for transforming a stereographic projection into an azimuthal equidistant. The existing graticule intersection, being stereographic, is noted to be 22.5° from the origin on the S scale. Its new location, on the equidistant projection, is along the same straight edge at 22.5° on the E scale. In the case shown, a scale reduction as well as a projection transformation is being accomplished.

The graticule intersections, like the cities, also lie in their correct directions from London on each of the oblique azimuthal projections centered there. This is true regardless of the scale adopted in any of the projections.

The azimuthal characteristic may be used to transform an oblique stereographic projection into any other azimuthal projection at the same or a different scale. It also is used in photogrammetry in the radial-line plotters, or slotted-templet method, wherein images on perspective photographs are corrected radially to their orthographic positions.

8-6 METHOD

The oblique stereographic is the obvious choice for the transformation process because it may be constructed with precision by compass, as shown in Secs. 8-1 and 8-3. If an oblique azimuthal equidistant projection, centered on the same spot as the stereographic, is to be created by transformation, the first step is to construct a scale with two sets of divisions as shown in Fig. 8-5. It is advantageous to plot them on opposite faces of Mylar drafting film or in two different colors to avoid confusion. The figure uses a letter code plus a slight offset. Both scales are in degrees. The divisions correspond to the radii used in drawing the parallels in the polar cases. The process is explained in the caption of the figure. Ideally, the work is done on a light table so that the stereographic graticule can be seen easily through the new sheet of film or paper. After a series of points on a particular meridian or parallel have been transformed, the line should be drawn with a flexible curve and numbered.

If the stereographic being used is limited to a hemisphere, the equidistant will initially be likewise limited. It can be expanded to cover the whole world, however. As an example of the procedure, assume that the point being transformed in Fig. 8-5 is at 60°N and 100°E. The antipodal point (60°S, 80°W) may be plotted at 180° on the E scale shown if the scale is laid with zero at "new" and with the straight edge passing exactly over the origin. During this operation, the stereographic graticule is not needed and may be removed.

More direct ways of constructing the equidistant projection are described in the following section.

OBLIQUE AZIMUTHAL EQUIDISTANT PROJECTION

8-7 CONSTRUCTION

The oblique azimuthal equidistant projection, by definition, will show each graticule intersection (and every other point) at its correct distance and azimuth from the origin. It is possible to use spherical trigonometry to calculate such polar coordinates for each intersection needed. They may then be plotted by scale and protractor. Finally, the meridians and parallels may be drawn in with the aid of a flexible curve. An example is shown in Fig. 7-12.

An improvement on the use of polar coordinates is to convert them to rectangular coordinates before plotting. Equations can be written for the x and y coordinates just as was done for the stereographic cases (Secs. 8-1 and 8-3). They may be solved by programmable pocket calculator as shown in Appendix A-4 or by electronic computer. In the latter case a plotting table may be printed out.

This idea was carried one step further in Fig. 8-6. A computer program was used not only to print the coordinates but also to plot them and to join the points with short line segments to closely approximate the curved parallels and meridians.

8-8 APPLICATIONS

The azimuthal equidistant projection is useful at airports, missile launching facilities, seismograph stations, for studies of radio transmission, and the like. In each case the map is centered at the location of the airport or scientific facility.

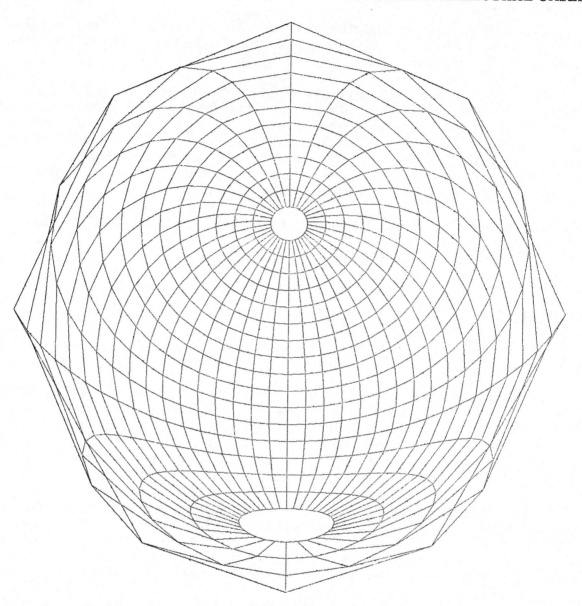

FIG. 8-6. Azimuthal equidistant projection centered on Pittsburgh, Pa. (ϕ = 40.3°N, λ = 80°W). Drawn by a computer and flat-bed plotter. Line segments shown are 10° in length. (Program written by Dr. O. O. Ayeni.)

OBLIQUE AZIMUTHAL EQUAL-AREA PROJECTION

8-9 CONSTRUCTION

The oblique azimuthal equal-area projection may be constructed by transformation like any of the other azimuthals (see Sec. 8-5). Also special tables may be used if available. Reference 2 includes a table set up for an origin on the 40th parallel and another for an origin on the Equator. Figure 8-7 shows the equatorial case

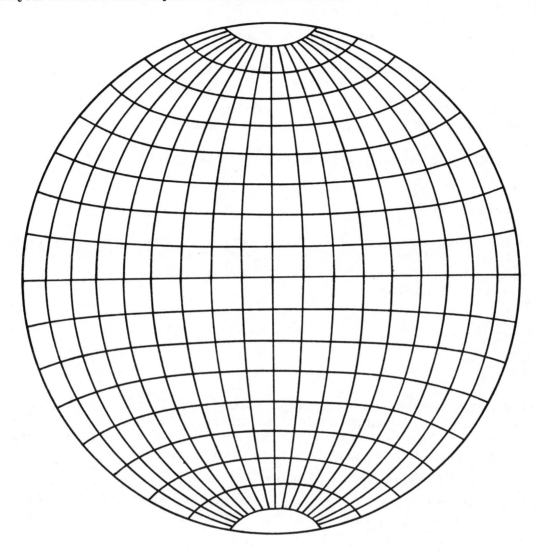

FIG 8-7 Lambert azimuthal equal-area projection, equatorial case with 10° interval. (Drawn by Thomas R. Lawther.)

plotted from these tables. (It is not intended that this case should be considered oblique, of course.)

Rectangular coordinates also may be obtained from the following equations [3]:

$$x = R \sqrt{2} \; \frac{\cos \phi \sin \lambda}{(1 + \sin \phi_0 \sin \phi + \cos \phi_0 \cos \phi \cos \lambda)^{1/2}} \qquad (6)$$

$$y = R \sqrt{2} \ \frac{(\cos \phi_O \sin \phi - \sin \phi_O \cos \phi \cos \lambda)}{(1 + \sin \phi_O \sin \phi + \cos \phi_O \cos \phi \cos \lambda)^{1/2}} \qquad (7)$$

For the equatorial case, $\phi_O = 0°$ and the equations are simplified. A pocket calculator program for Eqs. (6) and (7) is given in Appendix A-4.

A graphic solution exists only for the polar case. See Fig. 4-4.

8-10 APPLICATIONS

The azimuthal equal-area projection is widely used in atlases (Hammond, Times, Rand McNally, and Goode's), often for large regions such as Eurasia and North America. Some atlases refer to it as the Lambert zenithal equal-area or Lambert azimuthal equal-area.

PROBLEMS

8-1 Construct an equatorial stereographic projection for the Western Hemisphere showing meridians and parallels at 15° intervals. Use a circle with a radius r = 15 cm. State the RF scale. Save the drawing for comparison with the globular projection of the same size in Prob. 10-1.

   Partial Answer: RF = 1:84,930,000

8-2 Construct an oblique stereographic projection of a hemisphere by compass and scale as shown in the right half of Fig. 8-3. Center it on Melbourne, Australia ($\phi = 37.5°S$, $\lambda = 145°E$), use R = 4.0 cm and an interval of 30°.

   Partial Answer: For 85°E, K is at x = 5.82, y = 6.14

8-3 Verify mapping Eqs. (4) and (5) for plotting the oblique stereographic by x and y coordinates. Do this by solving them for $\phi = 60°S$, $\lambda = 175°E$, $\phi_O = 37.5°S$, and R = 4.0 cm; then scaling the corresponding values from your drawing for Prob. 8-2.

   Partial Answer: x = 1.069 cm

8-4 Use the HP-25 program given for the oblique stereographic (Appendix A-2) and plot a graticule like that of Prob. 8-2.

8-5 Refer to Sec. 8-5. Calculate the central angle between London and Rome (arc distance D in degrees, assuming the earth to be spherical). Find the map distance to Rome as it would appear on the following oblique projections centered on London. For London, $\phi = 51°$, $\lambda = 0°$; for Rome, $\phi = 41°$, $\lambda = 12°E$.

   a. Orthographic, 1:7,000,000
   b. Gnomonic, 1:7,000,000

    c. Azimuthal equal-area, 1:7,000,000
    d. Azimuthal equidistant, 1:7,000,000

    <u>Partial Answer:</u> b = 20.98 cm

8-6 Make a scale similar to the one shown in Fig. 8-5. For the stereographic scale assume that the radius of the generating globe is R = 18 cm. Assume that the transformation is to an oblique orthographic, R = 30 cm. Use a strip of paper or film about 38 cm in length and show 10° divisions. In practice, smaller divisions would be helpful.

REFERENCES

1. Paul D. Thomas, <u>Conformal Projections in Geodesy and Cartography</u>, Special Publication No. 251, U.S. Coast and Geodetic Survey (now National Geodetic Survey), Washington, D.C., 1952.

2. C. H. Deetz and O. S. Adams, <u>Elements of Map Projection</u>, Special Publication No. 68, U.S. Coast and Geodetic Survey (now National Geodetic Survey), Washington, D.C., 1944.

3. John P. Snyder, correspondence with the author, 1977.

9

OVAL PROJECTIONS

This chapter discusses two projections which show all of the globe within an oval boundary (actually an ellipse).   It also covers a variation in which one of the oval projections is combined with the sinusoidal projection (Sec. 5-1).   All of these projections are equal-area.

THE MOLLWEIDE PROJECTION

9-1 CONSTRUCTION

The first step in constructing a Mollweide projection for the whole world is to draw a circle of convenient size to represent one hemisphere.   The area of the circle is πr^2 and it represents a hemisphere with an area of $2\pi R^2$.   Thus the radius used to draw the circle (r) is related to the size of the generating globe as follows:

$$\pi r^2 = 2\pi R^2$$

$$r = R \sqrt{2}$$

The Equator is then added, as the horizontal diameter of the circle.

To make the figure twice as large in area so that it can represent both hemispheres, it is doubled in width, becoming an ellipse (see Fig. 9-1).   Other ellipses are added at equal intervals to represent meridians cutting off equal areas, as they do on the globe.   With the total area equal to that of the sphere $(4\pi R^2)$ and the meridians now dividing it correctly, the only matter remaining is to place the parallels so that they divide the area correctly too.

106

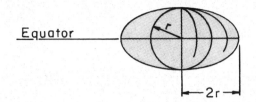

FIG 9-1 Initial steps in the construction of a Mollweide projection.

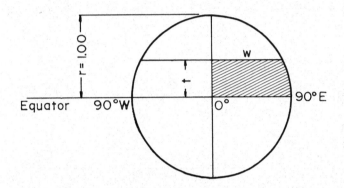

FIG 9-2 Locating the parallels on a Mollweide projection. The equation of the circle is $x^2 + y^2 = 1.00$ or $x = (1 - y^2)^{1/2}$.

In Fig. 9-2, the circle that is equal in area to a hemisphere is given a radius of r = 1.00 unit. The radius of the globe, therefore, is R = 0.7071 unit. The problem is to find values of t for plotting parallels such as 10°, 20°, etc., such that correct areas are shown.

As was shown in Fig. 1-5, the area on a globe between the Equator and a parallel is

$$2\pi R^2 \sin \phi$$

One-quarter of that quantity will be represented on the map by the shaded area in Fig. 9-2. The global area, therefore, is

$$\frac{\pi R^2 \sin \phi}{2}$$

The map area in the figure is

$$\int_0^t (1 - y^2)^{1/2} \, dy$$

When these are equated and the integral is evaluated, the following equation results:

$$\pi R^2 \sin \phi = \arcsin t + t(1 - t^2)^{1/2}$$

$$1.57080 \sin \phi = \arcsin t + t(1 - t^2)^{1/2}$$

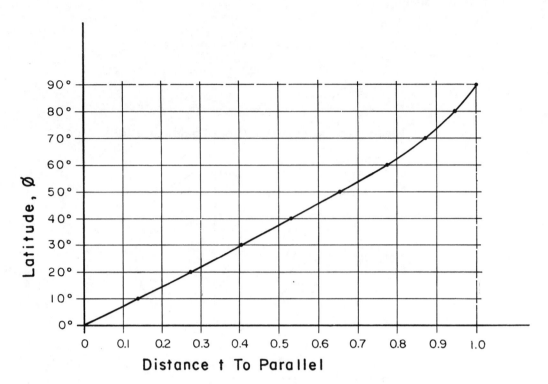

FIG 9-3 Graphic way of obtaining t in a Mollweide projection for any value of ϕ.

One way to solve this equation is to substitute values of t such as 0.10, 0.20, and 0.30 into the right side of the equation and solve it for ϕ. Then plot a graph as shown in Fig. 9-3. Values of t for any even value of ϕ may be read from the graph. (In doing the computation, arcsin t must be in radians.) The result is Table 9-1, which may be used for any selected value of r. The third column gives the length of each parallel, from the central meridian to the circle, called w in Fig. 9-2. In addition, w could be found from $w = (1 - t^2)^{1/2}$. It is not essential for drawing the graticule, as will be seen in the example to follow.

Example 9-1

Plan a graticule for a Mollweide projection of the world to fit inside of a 35 $\times$ 45 cm area. Fig. 9-4a shows such a map.

Solution

On the Mollweide projection the Equator is twice as long as the central meridian (see Fig. 9-1). Thus, in this example, the Equator will be 45.0 cm in length and the central meridian will be 22.5 cm.

Begin the construction by drawing a circle of radius r = 11.25 cm in the center of the sheet. This circle has the same area as the hemisphere of a globe with radius R.

$$R = \frac{r}{\sqrt{2}} = \frac{11.25}{\sqrt{2}} = 7.955 \text{ cm}$$

TABLE 9-1

Mollweide Projection.   Distances t and w for a Map Where r = 1. *

ϕ	t	w	ϕ	t	w
0°	0.0000	1.0000	50°	0.6512	0.7589
5°	0.0685	0.9976	55°	0.7080	0.7062
10°	0.1368	0.9906	60°	0.7624	0.6471
15°	0.2047	0.9788	65°	0.8138	0.5811
20°	0.2720	0.9623	70°	0.8619	0.5071
25°	0.3385	0.9410	75°	0.9060	0.4232
30°	0.4040	0.9148	80°	0.9454	0.3259
35°	0.4682	0.8836	85°	0.9784	0.2068
40°	0.5310	0.8474	90°	1.0000	0.0000
45°	0.5920	0.8059			

*See Fig. 9-2. (Based on Ref. 1.)

The RF scale is 7.955/637,000,000 = 1:80,080,000.   Parallels are drawn at t distances such as the following:

For ϕ = 10°, t = (0.1368)(11.25) = 1.54 cm

For ϕ = 20°, t = (0.2720)(11.25) = 3.06 cm

Each parallel is drawn to the circle and then doubled in length on both sides of the central meridian. If desired, the distances left and right may be calculated from Table 9-1 as follows:

For ϕ = 10°, distance = 2(0.9906)(11.25) = 22.29 cm

For ϕ = 20°, distance = 2(0.9623)(11.25) = 21.65 cm

The parallels are divided into equal intervals to locate the meridian crossings. This may be done by dividers or mathematically. Smooth curves are then drawn to form the meridians. Because the meridians are ellipses, it is also possible to construct them using a mechanical device called an ellipsograph. See Fig. 9-5.

Because the Mollweide has straight parallels and curved meridians, it is classified as pseudocylindrical. The sinusoidal (Sec. 5-1) is another example. There are, in fact, about 80 such projections [2]. None of the others will be covered in this book.

The Mollweide was developed in 1805 by Carl B. Mollweide.

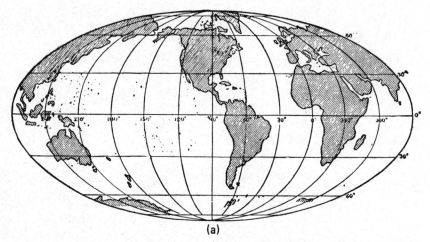

(a)

FIG 9-4a Mollweide projection of the world.   (From Ref. 1.)

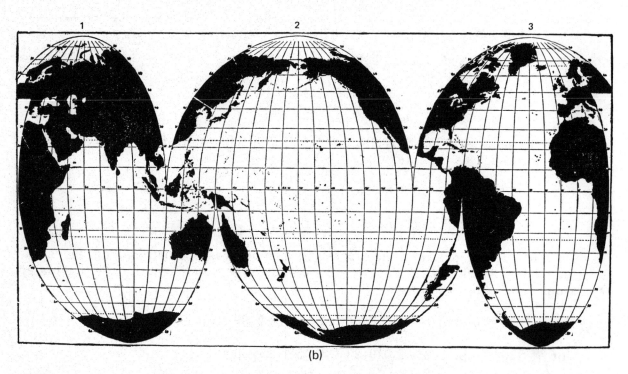

(b)

FIG 9-4b Interrupted Mollweide projection. Central meridians were chosen so that oceans would be featured and the interruptions would occur in the continents. (From Prof. J. Paul Goode, copyright by The University of Chicago, Department of Geography.)

9-2 APPLICATIONS

The Mollweide projection, like the sinusoidal, is useful for showing distributions (population, for example) for the whole world or a large part of it such as a hemisphere. In comparison to the sinusoidal, it offers a more rounded (oval) look near

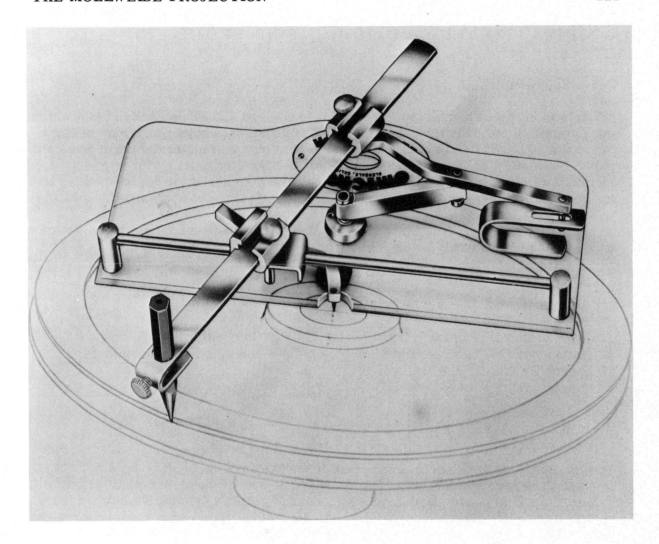

FIG 9-5 An ellipsograph, a device for constructing accurate ellipses. (Courtesy of The Omicron Company, Burbank, Cal.)

the poles. Countries located near the east or west edges of the map are badly distorted in shape, as evidenced by the oblique angles at which meridians and parallels meet. The shapes can be improved by "interrupting" or recentering the projection for each continent or ocean as was shown in the case of the sinusoidal (Fig. 5-3). (See Fig. 9-4b.)

In offering an equal-area view of the entire world at a glance, this and some other projections have an advantage over a globe. An observer can never see all of a globe at once.

The next two projections in this chapter offer slight advantages over the Mollweide, as will be noted.

GOODE'S INTERRUPTED HOMOLOSINE PROJECTION

9-3 DESCRIPTION

Goode's interrupted homolosine projection combines the sinusoidal and Mollweide projections. It involves the construction of a sinusoidal projection from the Equator to the 40th parallels, north and south. From that latitude to the north and south poles, the Mollweide is used (see Fig. 9-6). It offers the good features of both projections (equally spaced parallels, all of which are standard, in the sinusoidal; and the rounded look near the poles in the Mollweide).

Actually, when $\phi = 40°$ the widths of the two projections do not match exactly. If R = 1, the widths would be as follows:

Sinusoidal: $2\pi R \cos 40° = 4.813$

Mollweide: $4w \sqrt{2} = 4(0.8474) \sqrt{2} = 4.794$

The two projections actually match at $40°44'12''$ according to the originator, J. Paul Goode (1923). The name homolosine is a combination of homolographic, an alternate name for the Mollweide, and sinusoidal.

HAMMER – AITOFF PROJECTION

The Hammer-Aitoff projection was first described in 1892 by Dr. Hammer, a German professor of surveying, but until 1952 it was mistakenly called the Aitoff projection. This has caused confusion because Aitoff's name also is used (correctly) in reference to an older projection from which Hammer developed his idea. Currently, the name used is either Hammer-Aitoff or simply Hammer's projection. It is equal-area whereas Aitoff's original projection (of 1889) was not.

A world map on this projection is bounded by an ellipse of the same size and shape as is used in a Mollweide projection. Since both projections also are equal-area, they look somewhat alike. The Hammer-Aitoff projection, however, has curved parallels and thus is not classified as pseudocylindrical. The parallels meet the meridians at less oblique angles than is the case in the Mollweide (see Fig. 9-7).

9-4 CONSTRUCTION

This projection is derived from the azimuthal equal-area projection described in Secs. 4-5 and 8-9. The polar case, outlined in Sec. 4-5, was seen to cover a hemisphere in a circle of radius $r = 2R \sin 45°$ or $1.4142R$. An oblique or equatorial case (not centered on the pole) also covers a hemisphere in a circle of that size (see Fig. 8-7). Hammer realized that he could do the same thing to the equatorial azimuthal equal-area projection of a hemisphere that was done to the Mollweide projection of a hemisphere. He could double the x coordinates and obtain an ellipse that would contain an area equal to that of both hemispheres. If one has a table for plotting the equatorial azimuthal equal-area projection by coordinates, it is easy to set up a table for plotting a Hammer-Aitoff projection. To obtain an x

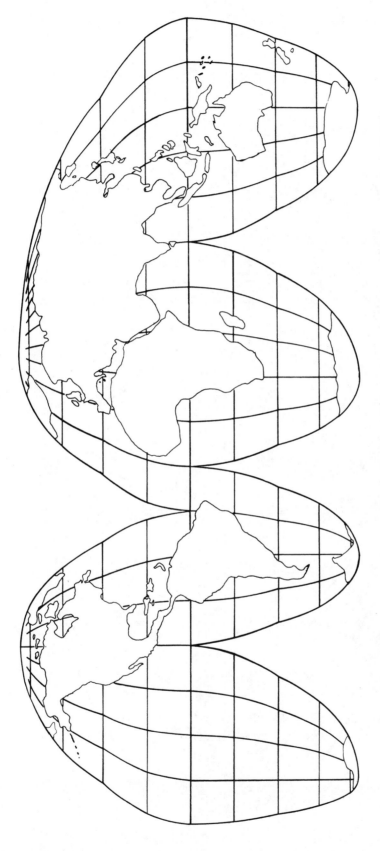

FIG 9-6 Goode's interrupted homolosine projection. (Copyright by The University of Chicago, Department of Geography; redrawn by Michael Compton.)

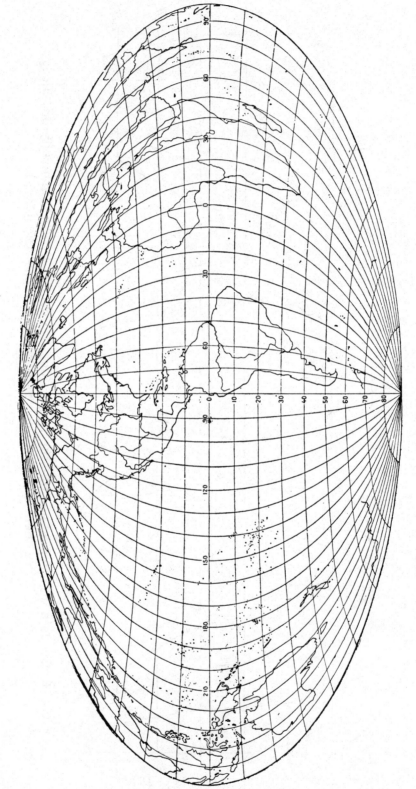

FIG 9-7 Hammer–Aitoff projection. (From Ref. 1.)

coordinate for plotting $\phi = 20°$, $\lambda = 40°$ on a Hammer-Aitoff projection, where the central meridian is at $\lambda = 0°$, one can double the value listed for $\phi = 20°$, $\lambda = 20°$. The y coordinate for $\phi = 20°$, $\lambda = 20°$ is used without change. Reference 1 contains both tables.

Coordinates for plotting a Hammer-Aitoff projection also may be calculated from the following equations [3]:

$$x = \frac{2\sqrt{2}R \cos \phi \sin 1/2 \lambda}{(1 + \cos \phi \cos 1/2 \lambda)^{1/2}} \tag{1}$$

$$y = \frac{\sqrt{2}R \sin \phi}{(1 + \cos \phi \cos 1/2 \lambda)^{1/2}} \tag{2}$$

These equations, having the same denominator, are easily handled by a computer or programmable pocket calculator.

Appendix A-3 shows a program for an HP-25 pocket calculator. It evaluates a quantity K which is common to both equations.

$$K = \frac{\sqrt{2}R}{(1 + \cos \phi \cos 1/2 \lambda)^{1/2}} \tag{3}$$

Then it finds x and y

$$x = 2K \cos \phi \sin 1/2 \lambda \tag{4}$$

$$y = k \sin \phi \tag{5}$$

These may be plotted immediately, in all four quadrants, or tabulated as shown in the following example.

Example 9-2

Find x and y coordinates for plotting a Hammer-Aitoff projection if the radius of the generating globe R = 10 cm.

Solution

Using the equations given, a plotting table may be set up as follows:

ϕ	Longitude from Central Meridian								(cont. to 180°)
	0°		15°		30°		45°		
	x	y	x	y	x	y	x	y	
0°	0.00	0.00	2.62	0.00	5.22	0.00	7.80	0.00	
15°	0.00	2.61	2.55	2.62	5.09	2.63	7.60	2.66	
30°	0.00	5.18	2.35	5.19	4.68	5.22	6.99	5.27	
45°	0.00	7.65	2.00	7.67	3.99	7.71	5.95	7.78	
60°	0.00	10.00	1.51	10.01	3.01	10.06	4.48	10.13	
75°	0.00	12.18	0.85	12.19	1.69	12.22	2.52	12.27	

A 10° interval was used in Fig. 9-7. The pole may be tabulated if desired (x = 0, y = 1.4142R).

9-5 APPLICATIONS

The projection has a realistic look, being equal-area, having curved parallels, and having its meridians and parallels meeting at less oblique angles than on some other oval projections. Because it shows the world in one connected whole, it is a good choice for showing physical geography as well as distributions. It has been used to represent the celestial sphere in astronomy where it shows the relative distribution of the stars in the different parts of the heavens, and also is seen in European atlases.

PROBLEMS

9-1 On a Mollweide projection with R = 10 cm, find the x and y coordinates of a point lying 30° north of the Equator and 100° east of the central meridian.

  Partial Answer: x = 14.37 cm

9-2 Solve Prob. 9-1 for a Hammer-Aitoff projection.

  Partial Answer: x = 15.04 cm

9-3 Find the scale factor along the 50th parallel of a Mollweide projection.

  Answer: Scale factor = 1.063

9-4 How many times larger in <u>area</u> (square units) is the surface of the earth than the surface of the ellipse in Probs. 9-1 and 9-2?

  Answer: 4.059×10^{15}

9-5 Roughly verify your answer to Prob. 9-4 by considering the quadrangle bounded by 20°N, 25°N, 30°E, and 35°E. Find its area on the Mollweide where it is roughly a trapezoid and on the earth where it is 5/360 of the area of a zone as found in Example 1-2.

  Partial Answer: Trapezoid = 0.70295 cm^2

9-6 Calculate a Mollweide projection for Africa using λ = 20°E for the central meridian. Use R = 15.28 cm and assume that the region is bounded by 20°W, 50°E, 35°S, and 40°N. (These are the same data as were used in Example 2-1 for the cylindrical equidistant and Example 5-1 for the sinusoidal projection.)

  Partial Answer: Dist. to 20°W on 40th = 8.14 cm

9-7 Complete the table given in Example 9-2 for a Hammer-Aitoff projection, R = 10 cm. Draw it, if assigned, noting that the bounding ellipse may be plotted without using the coordinates if an ellipsograph is available.

9-8 Draw a Mollweide projection for the whole world using R = 10 cm. Use drafting film or tracing vellum so that it may be placed over the Hammer-Aitoff projection of Prob. 9-7 for comparison.

  Partial Answer: Length of Equator = 56.56 cm

9-9 Use the relationship that exists between an equatorial azimuthal equal-area projection and a Hammer-Aitoff projection to calculate a graticule for Africa using the former. See Prob. 9-6 for the bounding parallels and meridians, and for R. Use Eqs. (4) and (5) in this chapter. The x coordinates for λ = 50°E, which lies 30° east of the central meridian, will be equal to half of the values found for a meridian 60° east of it on a Hammer-Aitoff projection. (This azimuthal projection, centered on the Equator, is indeed used for Africa in Goode's School Atlas, published by Rand McNally.)

   Partial Answer: for NE corner, x = 6.42 cm, y = 10.77 cm

REFERENCES

1. C. H. Deetz and O. S. Adams, Elements of Map Projection, Special Publication No. 68, U.S. Coast and Geodetic Survey (now National Geodetic Survey), Washington, D.C., 1944.

2. John P. Snyder, The American Cartographer, 4 (1): 59-81 (1977).

3. D. H. Maling, Coordinate Systems and Map Projections, International Publication Service, New York, 1974.

10

OTHER IMPORTANT PROJECTIONS

This chapter covers briefly the globular projection and then, in greater detail, the transverse Mercator projection. The latter is used in the state plane coordinate system in many states as well as in the Universal Transverse Mercator (military grid) system. The discussion of the military grid system necessarily includes the Universal Polar Stereographic projection as well.

GLOBULAR PROJECTION

10-1 CONSTRUCTION

The globular projection, illustrated in Fig. 10-1, is neither conformal nor equal-area and is not designed in terms of standard lines or any sort of cylinder, cone, or plane. It is constructed very directly, using a compass for all curved lines, and it is used only for hemispheres, as shown.

A circle is drawn to enclose a hemisphere. The horizontal diameter is the Equator and the vertical diameter is the central meridian. These diameters and the circumference are each divided into equal parts and circles are drawn as defined by three points. In the case of the meridians, of course, two of the points are the poles.

10-2 APPLICATIONS

Many years ago the globular projection was often used in atlases to show a hemisphere. Both shapes and areas are well represented. The globular projection resembles the equatorial stereographic (Sec. 8-1) in the sense that all meridians and parallels are circular.

TRANSVERSE MERCATOR

In the familiar Mercator projection (Sec. 6-1), the Equator is the only standard
line. The meridians are perpendicular to it and spaced at equal intervals. The
parallels are straight lines drawn parallel to the Equator and spaced so that the
graticule is conformal. The poles cannot be shown because they would be at infinity. The projection was classified as cylindrical, although it is a mathematical
creation and is not geometrically projected to a cylinder.

The transverse Mercator is a variation in which the central meridian is the
standard line (not the Equator). The "cylinder" is turned 90°, becoming transverse to the earth's axis rather than coaxial with it. The projection remains conformal and the meridians and parallels continue to meet at right angles, but the
pattern is not rectangular (see Fig. 10-2). Like the polyconic (Sec. 5-5), the result is an unfamiliar map of no great value if used for a large portion of the earth.
It does not show loxodromes as straight lines as does the conventional Mercator.
Its great value comes in its application to limited regions which lie close to a central meridian. It is important in surveying and military applications, but not for
drawing a map of the world or a hemisphere.

10-3 CONSTRUCTION

Because the transverse Mercator is used only for limited areas, the scale of the
maps generally is relatively large, such as 1:25,000. In the case of surveying
applications, the field measurements may be used at full scale (not just for plotting on a map). For both of these applications, the earth cannot be taken to be
spherical. A spheroid such as Clarke's of 1866 must be adopted. On a spheroid,
the central meridian is an ellipse rather than a circle, and the cylinder also is
elliptical. The computations for plotting such a graticule are beyond the scope
of this book. In fact, the problem is more complicated than suggested to this
point because the central meridian is usually assigned a scale factor smaller
than 1.000 (as was done with the Equator in the cylindrical equidistant with two
standard parallels, Sec. 3-1). Thus there are two standard lines neither of which
is a meridian.

Special tables have been prepared, one of which will be illustrated here and
discussed in Sec. 10-6.

10-4 APPLICATIONS

Two important applications will be described. One is in the state plane coordinate
system established by the U.S. Coast and Geodetic Survey (now the National Geodetic Survey), and the other is in the UTM (Universal Transverse Mercator) grid
pioneered by the U.S. Army and now widely adopted in other countries as well.

(a)

FIG 10-1 Comparison of globular projection (left) and stereographic projection. The central meridians used here are not the same (90° and 80°). (Drawings by Tim Witter and Douglas Brehm.)

10-5 THE STATE PLANE COORDINATE SYSTEM

The state plane coordinate system was designed to render geodetic control stations of greater value to highway engineers, land surveyors, and others who are accustomed to using rectangular coordinates (x and y) rather than geographic coordinates (ϕ and λ). In 1933, projections were designed for each state. The

(b)

larger states were divided into several zones because it was decided that scale factors should be very close to 1. 0000 (actually between 0. 99990 and 1. 00010).   The
Lambert conformal conic projection was chosen for those states, or zones, which
were long in the east-west direction such as Pennsylvania.   The transverse Mercator was chosen for Indiana and other states or zones which extended the other
way, along meridians.   A rectangular grid was superimposed upon each projection.   The central meridian was assigned a numerically large x coordinate, such
as 2,000,000 ft, so that any survey station would have a positive x coordinate.

State plane coordinates offer great advantages in major engineering projects
such as highway design and construction, and in photogrammetric mapping, property surveying, and regional planning.   They provide a common basis for all

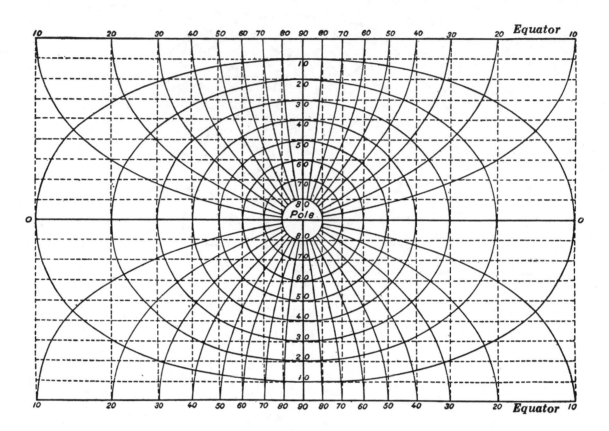

FIG 10-2 Transverse Mercator projection centered on pole (solid lines). Also
shown, in dashed lines, is a conventional Mercator projection turned sideways.
(From Ref. 1.)

horizontal control surveys. Similar systems exist in other countries. Conformal
projections should be used as was mentioned in Sec. 1-12. The author in 1975 published a manual called <u>Simplified Tables for the Pennsylvania Coordinate System</u>,
which covers the Lambert system in particular. Because tables for state plane coordinates are used for surveying computations as well as for plotting maps, they
are published for an RF scale of 1:1 (full scale), unlike Tables 6-1 and 6-2, for
example.

10-6 THE UTM GRID

The Universal Transverse Mercator is one of several grids that have been devised
to simplify the problems of giving directions, distances, and positions in military
operations.

The first real need for a military grid came in World War I when the French
developed artillery with a range of 5 miles (8 km), far beyond the gunner's vision.
A new firing technique was needed, based on the known locations of gun and target.
The computations for direction and distance had to be simple, rapid, and accurate.

They would be too involved if based on latitude and longitude. The French adopted rectangular grids superimposed upon the Lambert conformal conic projection. After the war, the Germans and Russians adopted transverse Mercator grids. Unfortunately, the United States established a grid based upon the polyconic, which is not conformal. The British developed several systems, in yards, for parts of their empire. By the end of World War II there were about 100 heterogeneous grids in use. At that time the U.S. Army Map Service reviewed the existing situation and devised the UTM grid, plus the Universal Polar Stereographic (UPS), mentioned in Sec. 6-6, to cover the whole world in a well-planned system using metric units.

The UTM and UPS grids have grown in importance in recent years. They are used in Europe and Russia for surveying purposes as well as for mapping and military purposes.

There are 60 UTM and two UPS zones which, together, cover the whole world. Each of the UTM zones covers 6° of longitude and extends from 80°S to 84°N. The UPS zones cover the two polar areas. Figure 10-3 shows that the UTM zones are numbered from 1 to 60 beginning at 180°W. Los Angeles, California, for example, is in zone 11, which is bounded by 120° and 114°W. Zone 15, as shown, is bounded by 96° and 90°W. All boundaries are divisible by 6. The zones are subdivided into quadrangles or "grid zone designations" covering 8° of latitude which are lettered as shown from C to X (with I and O omitted to avoid confusion with numbers). The top row of grid zone designations covers 12° of latitude instead of only 8°.

Figure 10-4 shows the rectangular grid (100,000-m squares) superimposed upon the projection of a zone. Each zone has its own central meridian along which the scale factor is 0.9996. The standard lines are grid lines, not meridians, and occur on each side of the origin at a distance of 180,000 m (see Fig. 10-5). The origin is assigned an easting (or x coordinate) of 500,000 m so that no point in the zone will have a negative coordinate. The assigned value is called a false easting. For northings, two false values are assigned. For points above the Equator, the Equator is given a northing of 0 m; for points below it the Equator is assigned a northing of 10,000,000 m.

Figure 10-6 shows a sample page from UTM grid tables published by the U.S. Army. The manual includes 162 similar pages giving the eastings and northings for the 7-1/2 minute intersections north of the Equator in any zone. (The zones are exactly alike.) To construct a UTM projection on paper, the network of grid lines is plotted first as a series of perfect squares at the chosen scale. The intersections of meridians and parallels are then plotted on the grid using the values in the tables. Where the map covers just 7-1/2 minutes at a scale of 1:25,000, it is sufficient to plot only the four corners and connect them with straight lines. The 2-1/2 minute ticks may be obtained by subdivision. At this scale the curvature of the parallels is too small to plot. (It is in the order of 0.08 mm.) For maps covering greater areas at smaller scales, curvature will appear. The size of the grid squares will be the same for any map of a given scale. Thus it is convenient to plot a master grid on stable material and make copies for future use. With electronic computers automatic plotting of both the grid and the graticule has become possible.

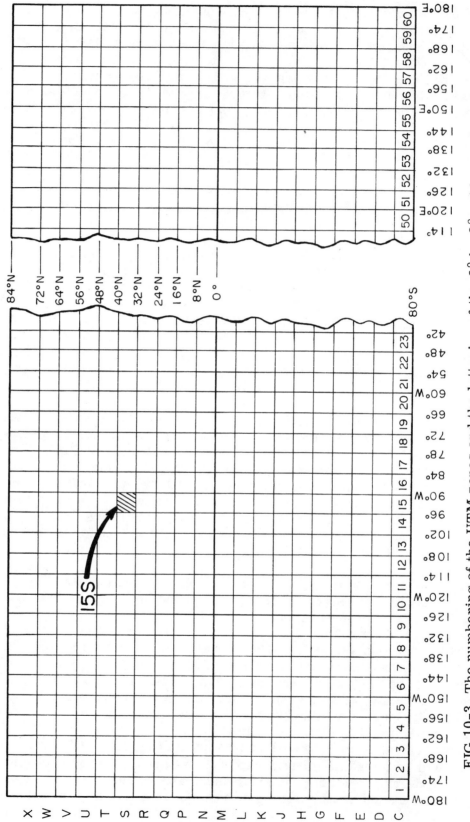

FIG 10-3 The numbering of the UTM zones and the lettering of the 6° by 8° grid zone designations. Row X covers 12° of latitude instead of 8°.

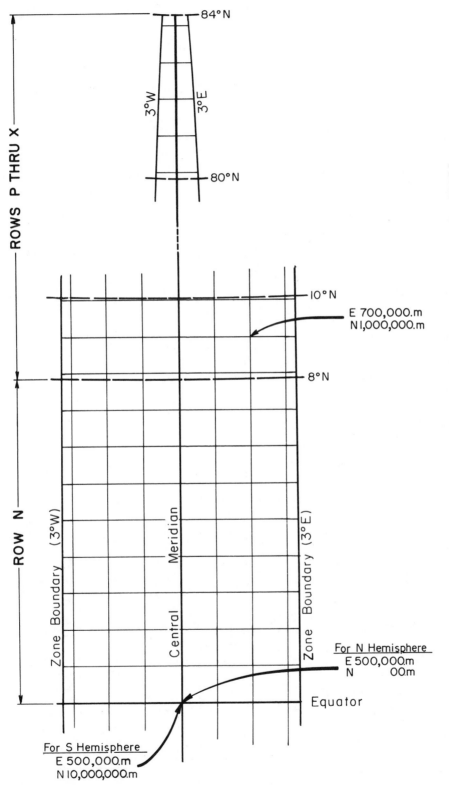

FIG 10-4 Rectangular grid superimposed on the projection of a UTM zone.

126 OTHER IMPORTANT PROJECTIONS

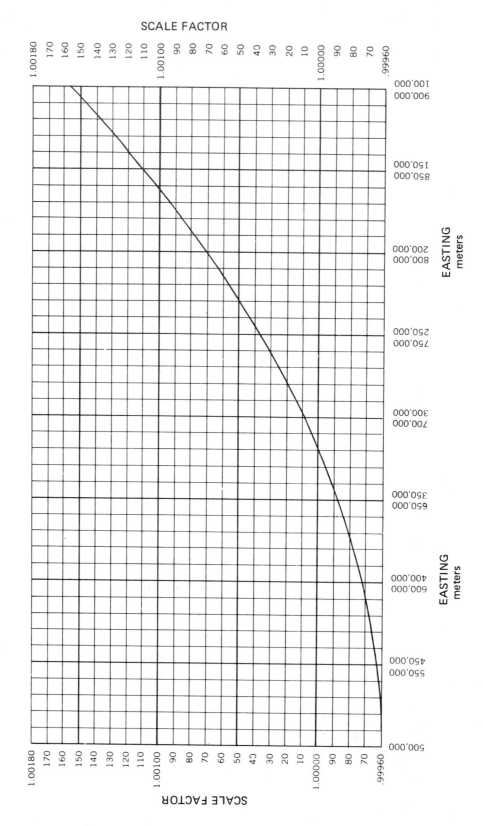

FIG 10-5 Variation of the scale factor across the width of a UTM zone and beyond. Standard lines, parallel to the central meridian, occur at 320,000 and 680,000 m E. (From Ref. 2.)

LATITUDE 38°00'00"

Δλ	West of C.M. E	East of C.M. E	N
0°00'00"	5 00,000.0	5 00,000.0	4,205,609.3
07 30	4 89,025.1	5 10,974.9	4,205,616.7
15 00	4 78,050.1	5 21,949.9	4,205,638.8
22 30	4 67,075.1	5 32,924.9	4,205,675.7
30 00	4 56,100.1	5 43,899.9	4,205,727.3
37 30	4 45,125.0	5 54,875.0	4,205,793.6
45 00	4 34,149.9	5 65,850.1	4,205,874.7
52 30	4 23,174.7	5 76,825.3	4,205,970.5
1 00 00	4 12,199.4	5 87,800.6	4,206,081.1
07 30	4 01,224.0	5 98,776.0	4,206,206.4
15 00	3 90,248.5	6 09,751.5	4,206,346.5
22 30	3 79,272.8	6 20,727.2	4,206,501.3
30 00	3 68,297.0	6 31,703.0	4,206,670.9
37 30	3 57,321.1	6 42,678.9	4,206,855.2
45 00	3 46,345.0	6 53,655.0	4,207,054.3
52 30	3 35,368.7	6 64,631.3	4,207,268.1
2 00 00	3 24,392.2	6 75,607.8	4,207,496.7
07 30	3 13,415.6	6 86,584.4	4,207,740.1
15 00	3 02,438.7	6 97,561.3	4,207,998.3
22 30	2 91,461.5	7 08,538.5	4,208,271.2
30 00	2 80,484.2	7 19,515.8	4,208,558.9
37 30	2 69,506.6	7 30,493.4	4,208,861.3
45 00	2 58,528.7	7 41,471.3	4,209,178.6
52 30	2 47,550.5	7 52,449.5	4,209,510.6
3 00 00	2 36,572.0	7 63,428.0	4,209,857.4
07 30	2 25,593.3	7 74,406.7	4,210,219.1
15 00	2 14,614.2	7 85,385.8	4,210,595.5
22 30	2 03,634.8	7 96,365.2	4,210,986.8
30 00	1 92,655.0	8 07,345.0	4,211,392.8

LATITUDE 38°15'00"

Δλ	West of C.M. E	East of C.M. E	N
0°00'00"	5 00,000.0	5 00,000.0	4,233,347.4
07 30	4 89,062.4	5 10,937.6	4,233,354.7
15 00	4 78,124.8	5 21,875.2	4,233,376.9
22 30	4 67,187.2	5 32,812.8	4,233,413.8
30 00	4 56,249.6	5 43,750.4	4,233,465.5
37 30	4 45,311.8	5 54,688.2	4,233,532.0
45 00	4 34,374.1	5 65,625.9	4,233,613.3
52 30	4 23,436.2	5 76,563.8	4,233,709.3
1 00 00	4 12,498.3	5 87,501.7	4,233,820.1
07 30	4 01,560.3	5 98,439.7	4,233,945.7
15 00	3 90,622.2	6 09,377.8	4,234,086.1
22 30	3 79,683.9	6 20,316.1	4,234,241.2
30 00	3 68,745.5	6 31,254.5	4,234,411.2
37 30	3 57,807.0	6 42,193.0	4,234,595.9
45 00	3 46,868.3	6 53,131.7	4,234,795.4
52 30	3 35,929.4	6 64,070.6	4,235,009.7
2 00 00	3 24,990.3	6 75,009.7	4,235,238.8
07 30	3 14,051.1	6 85,948.9	4,235,482.7
15 00	3 03,111.6	6 96,888.4	4,235,741.4
22 30	2 92,172.0	7 07,828.0	4,236,014.9
30 00	2 81,232.0	7 18,768.0	4,236,303.2
37 30	2 70,291.9	7 29,708.1	4,236,606.3
45 00	2 59,351.5	7 40,648.5	4,236,924.3
52 30	2 48,410.8	7 51,589.2	4,237,257.0
3 00 00	2 37,469.8	7 62,530.2	4,237,604.6
07 30	2 26,528.6	7 73,471.4	4,237,967.0
15 00	2 15,587.0	7 84,413.0	4,238,344.2
22 30	2 04,645.1	7 95,354.9	4,238,736.3
30 00	1 93,702.9	8 06,297.1	4,239,143.3

LATITUDE 38°07'30"

Δλ	West of C.M. E	East of C.M. E	N
0°00'00"	5 00,000.0	5 00,000.0	4,219,478.2
07 30	4 89,043.7	5 10,956.3	4,219,485.6
15 00	4 78,087.4	5 21,912.6	4,219,507.7
22 30	4 67,131.1	5 32,868.9	4,219,544.6
30 00	4 56,174.7	5 43,825.3	4,219,596.3
37 30	4 45,218.3	5 54,781.7	4,219,662.7
45 00	4 34,261.8	5 65,738.2	4,219,743.8
52 30	4 23,305.3	5 76,694.7	4,219,839.8
1 00 00	4 12,348.6	5 87,651.4	4,219,950.4
07 30	4 01,391.9	5 98,608.1	4,220,075.9
15 00	3 90,435.1	6 09,564.9	4,220,216.1
22 30	3 79,478.1	6 20,521.9	4,220,371.1
30 00	3 68,521.0	6 31,479.0	4,220,540.9
37 30	3 57,563.7	6 42,436.3	4,220,725.4
45 00	3 46,606.3	6 53,393.7	4,220,924.7
52 30	3 35,648.7	6 64,351.3	4,221,138.8
2 00 00	3 24,690.9	6 75,309.1	4,221,367.7
07 30	3 13,732.9	6 86,267.1	4,221,611.3
15 00	3 02,774.7	6 97,225.3	4,221,869.7
22 30	2 91,816.3	7 08,183.8	4,222,142.9
30 00	2 80,857.6	7 19,142.4	4,222,430.9
37 30	2 69,898.7	7 30,101.3	4,222,733.7
45 00	2 58,939.5	7 41,060.5	4,223,051.3
52 30	2 47,980.0	7 52,020.0	4,223,383.7
3 00 00	2 37,020.3	7 62,979.7	4,223,730.9
07 30	2 26,060.3	7 73,939.7	4,224,092.9
15 00	2 15,099.9	7 84,900.1	4,224,469.8
22 30	2 04,139.2	7 95,860.8	4,224,861.4
30 00	1 93,178.2	8 06,821.8	4,225,267.9

LATITUDE 38°22'30"

Δλ	West of C.M. E	East of C.M. E	N
0°00'00"	5 00,000.0	5 00,000.0	4,247,216.8
07 30	4 89,081.2	5 10,918.8	4,247,224.2
15 00	4 78,162.3	5 21,837.7	4,247,246.4
22 30	4 67,243.5	5 32,756.5	4,247,283.4
30 00	4 56,324.6	5 43,675.4	4,247,335.1
37 30	4 45,405.6	5 54,594.4	4,247,401.7
45 00	4 34,486.6	5 65,513.4	4,247,483.0
52 30	4 23,567.6	5 76,432.4	4,247,579.1
1 00 00	4 12,648.4	5 87,351.6	4,247,690.1
07 30	4 01,729.2	5 98,270.9	4,247,815.8
15 00	3 90,809.8	6 09,190.2	4,247,956.3
22 30	3 79,890.3	6 20,109.7	4,248,111.6
30 00	3 68,970.7	6 31,029.3	4,248,281.7
37 30	3 58,050.9	6 41,949.1	4,248,466.7
45 00	3 47,131.0	6 52,869.0	4,248,666.4
52 30	3 36,210.9	6 63,789.1	4,248,880.9
2 00 00	3 25,290.7	6 74,709.4	4,249,110.3
07 30	3 14,370.2	6 85,629.8	4,249,354.4
15 00	3 03,449.5	6 96,550.5	4,249,613.4
22 30	2 92,528.6	7 07,471.4	4,249,887.2
30 00	2 81,607.5	7 18,392.5	4,250,175.8
37 30	2 70,686.2	7 29,313.8	4,250,479.2
45 00	2 59,764.6	7 40,235.4	4,250,797.4
52 30	2 48,842.7	7 51,157.3	4,251,130.6
3 00 00	2 37,920.6	7 62,079.4	4,251,478.5
07 30	2 26,998.2	7 73,001.8	4,251,841.3
15 00	2 16,075.4	7 83,924.6	4,252,218.9
22 30	2 05,152.4	7 94,847.6	4,252,611.4
30 00	1 94,229.0	8 05,771.0	4,253,018.7

FIG 10-6 Sample page from UTM grid tables, U. S. Army Manual TM 5-241-11, Dec. 1959.

The tables may be used for the Southern Hemisphere by subtracting the northing shown from 10,000,000. It should be noted that the UTM system is based on different spheroids in various parts of the world. The values in Fig. 10-6 would apply exactly only where the Clarke spheroid of 1866 is in use, such as in the United States. The Army has published similar tables for other spheroids such as the Bessel and the International.

Example 10-1

A UTM grid has been plotted for an area bounded by 35°N, 40°N, 66°W, and 72°W. The scale is 1:500,000. In what zone does the area fall? If the 38th parallel is plotted on the grid, how much will it deviate from a straight line?

Solution

Figure 10-3 shows that the mapped area extends for the full width of zone 19. Figure 10-6 shows that the northings along the 38th parallel will vary from about 4,205,609 m at the central meridian (69°W) to about 4,209,857 m at the edges of the map, 3° east and west. The maximum deviation from a straight line drawn across the map will be

$$\frac{4,209,857 - 4,205,609}{500,000} = 0.0085 \text{ m or } 8.5 \text{ mm}$$

The location of any point within the 164° range of latitude covered by the UTM grid system may be uniquely defined by stating the zone number, the easting and northing, and which hemisphere it is in. The grid zone designation (letters from C to X) may be used in place of identifying the hemisphere.

10-7 THE UPS GRID

The polar areas are treated separately using a different conformal projection. A rectangular grid is superimposed on a polar stereographic projection as shown in Fig. 10-7. The origins (the poles) are assigned false eastings and false northings of 2,000,000 m. The orientation of the grid (the definition of grid north) was quite arbitrary, of course. The figure shows the orientation chosen in the case of each pole.

The polar grids are divided into two designations using the remaining letters of the alphabet. Points in areas B and Z have eastings greater than 2,000,000 m and have east longitudes. The reverse is true for areas A and Y as shown. The grids of these two polar zones are designed to overlap into areas C and X in the UTM zones. A point with a latitude of 79.8°S would fall outside of A and B but its location could be expressed in terms of the UPS grid if desired. The 60 UTM zones also overlap each other in a similar manner, as do those of the state plane coordinate system, Sec. 10-5. This overlap may be seen in Fig. 10-6 where $\Delta\lambda$ is tabulated to 3°30' rather than only to 3°00'.

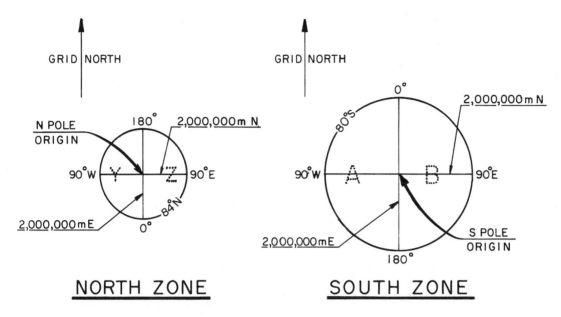

NORTH ZONE   SOUTH ZONE

FIG 10-7 The two zones of the UPS grid system showing false eastings and north-ings. Each zone contains two grid zone designations, namely, A and B in the south zone and Y and Z in the north zone. Points having west longitudes fall in A and Y, those with east longitudes fall in B and Z. Both zones originally ex-tended to the 80th parallel but the north zone now is bounded by the 84th parallel.

The UPS projection differs from the polar stereographic covered in Chap. 6. The latter was projected to a plane that was tangent at the pole whereas the UPS projection uses a "secant plane" which cuts the spheroid along a standard paral-lel, namely, 81°06'.

10-8 THE MILITARY GRID REFERENCE SYSTEM

The 62 grids just described are sufficient for identifying any point on earth. The lettered designations from A to Z are not essential to the coordinate systems. Using the numbers alone to express a position is said to be the civilian system [3]. Because it is numerical, the civilian system is readily handled by computers and data-processing systems. It does, however, involve a large number of digits. Northings, in particular, usually require nine digits if expressed to the nearest centimeter. To avoid such large numbers another system, called the military grid reference system, has been devised by military map makers.

The military grid reference system uses the lettered grid zone designations A through Z as previously described, plus additional letters. Each zone is divided into 100,000-m squares identified by two letters. Within a given square it becomes possible to give coordinates with respect to the southwest corner rather than only with respect to the origin on the Equator.

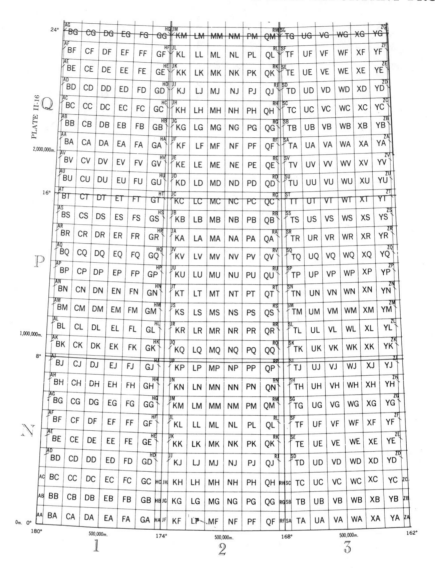

FIG 10-8 100,000-m square identifications for the military grid reference system. Parts of zones 1, 2, and 3 are shown. (Illustration by Defense Mapping Agency Topographic Center.)

The system of lettering to identify the 100,000-m squares is designed to keep squares carrying the same two-letter name as far apart as possible. Figure 10-8 shows the pattern for rows N, P, and Q of zones 1, 2, and 3. The first letter of the pair indicates the column; the second letter indicates the row. There is a square called BA in zone 1N as well as in zone 1Q. The complete designation of the latter square is 1QBA. The southwest corner of this square has an easting of 200,000 m and a northing of 2,000,000 m but the coordinates may be considered to be zero if the position of an interior point is given with respect to the square.

If the coordinates of a house in that square were scaled on a map as E 14,170 m and N 84,090 m and they were read to the nearest 10 meters, the conventional way to list the location is

1QBA14178409

in which punctuation and spaces are omitted. It is understood that eastings are given first, that trailing zeros are omitted, and that eastings and northings will both require the same number of digits. If the location was obtained less precisely, as E 14,200 and N 84,100, the designation would be

1QBA142841

in which it is understood that two trailing zeros are dropped. (Because square BA is 100,000 m on a side, it is known that the coordinates will always involve five digits before the decimal point.) Maps often show 10,000-m grid squares. In this example, the southwest corner of the square would be at 1QBA18.

If a large-scale map is available and a point can be located within one meter, there will be a string of 10 digits following the 1QBA designation. The UPS zones also are subdivided into 100,000-m squares (see Fig. 10-9).

The reference system described is useful in map-reading work, artillery firing, etc., but when the UTM (or UPS) grid is used for precise work, as in surveying, the unbroken string of digits would be inconvenient. Decimal fractions of a meter would be difficult to show. In this type of work the civilian system is used. The two coordinates are separated and decimal points are included. As stated earlier, the reference to the 100,000-m squares is omitted. As an example, assume that one corner of the house referred to in these paragraphs has been located by field surveys. Its position might be

Northern Hemisphere, zone 1

214,168.33 - 2,084,087.26

in which it is understood that the easting is given first.

10-9 PRINCIPAL DIGITS

Topographic maps, such as those of the U.S. Geological Survey, often show UTM grid ticks along the edge or "neatline." The grid lines may be drawn in by joining the grid ticks on opposite neatlines. In labeling the grid ticks it is customary to print two of the digits in larger type than the others and to omit the last three zeros except once on each neatline. For example, a map showing the house referred to in the previous section might have grid ticks labeled as follows:

2|2000m.E	2084000m.N
2	3	2085
2	4	2086
2	5	2087

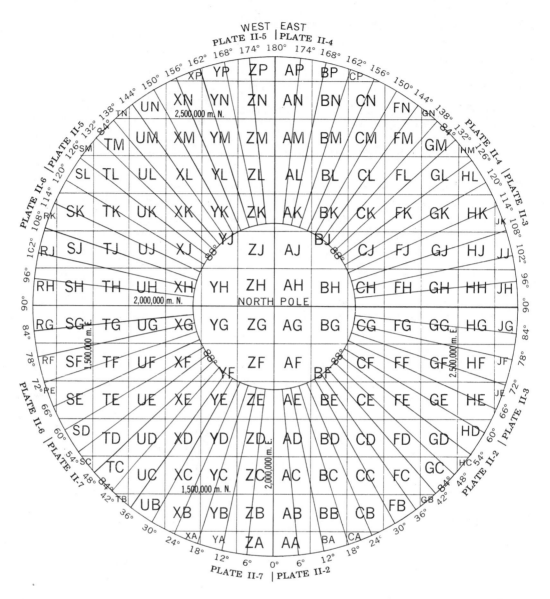

FIG 10-9 The two-letter identifications of the 100,000-m squares and partial squares in the north zone. (Illustration by Defense Mapping Agency Topographic Center.)

The principal digits are always in the thousands and ten-thousands places. On a smaller scale map showing only 10,000-m tick marks, only the first of the principal digits is shown. For the present example, they might be as follows:

210000m.E   2080000m.N
22   209
23   210

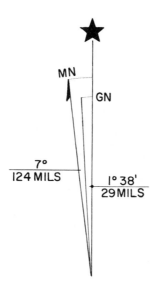

FIG 10-10 Typical declination diagram. The geographic meridian, indicated by
the star, converges toward the central meridian of the zone (in the Northern Hemisphere) while the grid meridian GN does not. The diagram also shows the magnetic meridian MN with a westward declination.

10-10 CONVERGENCE

The addition of a rectangular grid system on top of a map projection introduces a
new reference direction. In addition to geographic north, indicated by the graticule, there is a grid north. The angle between the two kinds of meridians is
called the grid convergence. Figure 10-10 shows a typical declination diagram
taken from the margin of a map. The star indicates geographic or "true" north,
GN indicates grid north, and the half-feathered arrow labeled MN indicates magnetic north for the year stated. The convergence shown (1°38') is called +, which
is the opposite of the sign convention used in the state plane coordinate system.
The diagram shown applies to a map in central Pennsylvania which lies west of
the central meridian. The magnetic declination shown (7°) is said to be a west
declination. In artillery work, angles are measured in mils (6,400 mils = 360°).

In a UTM zone, the numerical value of convergence is approximately

$$C = \Delta\lambda \sin \phi \tag{1}$$

where C is the grid convergence in degrees, minutes, or seconds; $\Delta\lambda$ is the difference in longitude between the location and the central meridian in the same units;
and ϕ is the latitude. The formula generally is correct to within 1 second.

The U.S. Army uses grid and true azimuths from north. (Geodesists use azimuths from south.)

Example 10-2

Find the true azimuth from north of a survey line from Station Axel to Station Rock if Axel is at 79°57'34"N and 90°01'22"W and the grid azimuth to Rock is 140°00'00".

Solution

The zone boundaries, shown in Fig. 10-3, are 90°W and 96°W. The central meridian is at 93° and Axel is east of it, close to the boundary. The geographic meridian will converge toward the central meridian (westward) while the grid meridian will be parallel. From Eq. (1),

C = (93°-90°01'22") sin 79°57'34" = 2°55'54"

The clockwise angle (azimuth) from the geographic meridian will be larger than 140°00'00", namely, 142°55'54".

Even though the convergence is large in the above example (it will rarely exceed 3°), the approximate formula gives the same result as was found from special tables in Ref. 4.

In a UPS zone, grid convergence at any point is equal to the longitude of the point. It varies from 0° to 180° because the geographic meridians are radial lines.

10-11 ZONE TO ZONE TRANSFORMATION

Surveying computations are much simpler if the plane coordinates of a UTM or UPS grid are used in place of the curvilinear coordinates ϕ and λ. When the survey extends across a zone boundary, however, some of the grid data may have to be converted into coordinates and azimuths expressed in terms of an adjacent zone. Special tables, as well as computer programs, are available for this purpose. (The same problem occurs in the state plane coordinate system or any other grid system involving several zones.) In many cases, surveying computations are done in terms of a single zone even though a few stations are in the nearby portion of the next zone.

Maps of areas close to a boundary often show grid lines for the major zone and grid ticks, in another color, for the overlapping zone.

PROBLEMS

10-1 Construct a globular projection of the Western Hemisphere using a circle with a radius of 15 cm. Use a 30° interval for the parallels and meridians. Compare it to the equatorial stereographic projection (Prob. 8-1).

10-2 Plot a 100,000-m square using a scale of 1 cm = 2.5 km. Add 10,000-m lines, making them dashed or lighter in weight than the main square. This is to be a 100,000-m square in zone 17 (Northern Hemisphere) of the UTM grid. Label the eastings from 700,000 m to 800,000 m, making the principal digits larger, as described in Sec. 10-9. Label the northings from 4,200,000 m to 4,300,000 m. Using Fig. 10-6, plot 15-minute intersections for ϕ = 38°00' and 38°15'. Add the parallels and meridians using straight lines for the latter. In which 6° by 8° grid zone designation does this area fall? Does the sheet as drawn extend into another zone? Consult an atlas and determine what major city lies less than a degree south of the southeast corner of your sheet.

10-3 In Prob. 10-2, one cannot determine the two-letter name for the 100,000-m square shown without a military map or more information on the system of lettering than is given in this text. Accept the fact that it is QN. Give the complete grid reference for a point at E 725,000 m, N 4,244,000 m, both coordinates being known to the nearest 1000 meters. Also give the reference for the southwest corner of the 10,000-m square in which it falls. Add the point to your drawing and call it Depot. By scaling, find the approximate ϕ and λ of Depot.

   Partial Answer: λ = 78°26'W

10-4 Find the approximate grid convergence at the Depot of Prob. 10-3. If a line has an azimuth from geographic north of 150°, what would its grid azimuth be?

10-5 Refer to an atlas and comment on the possibility of a house being located as described in Sec. 10-8.

10-6 Find the width of the map discussed in Example 10-1, Sec. 10-6.

10-7 Using Fig. 10-6, find the northing for ϕ = 38° S, λ = 153°E.

10-8 Plot the standard lines on Fig. 10-4 using dashed lines.

REFERENCES

1. C. H. Deetz and O. S. Adams, Elements of Map Projection, Special Publication No. 68, U.S. Coast and Geodetic Survey (now National Geodetic Survey), 5th Edition, Washington, D.C., 1944.

2. Universal Transverse Mercator Grid Tables, Clarke 1866 Spheroid, Volume 1, Technical Manual TM5-241-4/1, Dept. of the Army, Washington, D.C., 1958.

3. The Universal Transverse Mercator Grid, Department of Energy, Mines, and Resources, Surveys and Mapping Branch, Ottawa, Canada, 1969.

4. Artillery Survey, Technical Manual TM 6-200, Dept. of the Army, Washington, D.C., 1954.

11

PROJECTION OF GEOGRAPHIC FEATURES

The preceding chapters of this book are concerned with the projection of the graticule from a generating globe to a plane (usually a map but sometimes a rectangular grid system for use in surveying or military operations). The graticule is, of course, only a framework for the map itself. Nothing has been said about projecting the geographic features themselves. This chapter will deal with "map compilation," as it is called, and will include some comments about the plotting of the grid and/or graticule.

11-1 ORIGINAL MAPS AND DERIVED MAPS

Generally, a large-scale map, such as one intended for highway design, is based directly on field surveys and photogrammetry. Usually the framework involved is a state plane coordinate grid or some other rectangular system. It is an original or basic map based upon measurements of the terrain.

A small-scale map, showing an entire state or country, for example, is generally _derived_ from existing maps of larger scale. The topographic map series of a region is a common source of information for compilation at a smaller scale. Because of the smaller scale and broader coverage, it may be necessary to show meridians and parallels rather than only a local grid.

The purpose of the small-scale map may be to display some census data or other new information, but the coastlines, state boundaries, rivers, and other features shown as the background are not based on new measurements. In fact, one can buy "outline maps" of various parts of the world on which these derived coastlines have been compiled already. The subject of the final map may be hog production and the cartographic problem is how best to present the data on the outline map. This is the problem of "thematic mapping," not covered here.

An important aspect of deriving small-scale maps from large-scale (original) maps is the matter of _generalizing_. A map of Norway would not be made by a simple scale reduction of nautical charts. The islands, peninsulas, and inlets on the west coast (the fjords) would be shown in great detail for the mariner but in much less detail for the geography student. The irregularity would be indicated, but some little islands would have to be omitted. Generalization is the simplification or "characterizing by symbolization" which takes place in all mapping. Even on large-scale drawings such as house plans there will be generalization; a brick wall may be generalized as a smooth straight line. On a U.S. Geological Survey quadrangle map (1:24,000 or 1:25,000) all highways have the same width, all houses are squares, and driveways are omitted. There would be much less generalizing on a property survey of one parcel.

In deriving land features from existing maps it is important to work in the proper direction. Small-scale maps may be derived from large-scale maps, but not the reverse. One may not prepare nautical charts by enlarging from a small-scale map which was appropriately generalized. And surveyors cannot make a map with a 5-ft or 2-m contour interval by simply enlarging a U.S. Geological Survey quadrangle map and inserting contours at regular intervals.

CONSTRUCTING THE MAP

11-2 PLOTTING THE GRATICULE

There are five ways by which the rectangular grid and/or the graticule may be constructed. Obviously, the quality of the subsequent compilation will depend on the precision of this first operation.

Manual construction. It is possible to construct some of the graticules directly by scale and beam compass. Very often, however, a rectangular grid is necessary first. In fact, for large-scale maps the meridians and parallels may not be needed at all, and a UTM or other grid may be the only requirement (Sec. 1-1 mentioned that these are often called plans). On smaller scale maps the graticule intersections may be plotted from a grid by x and y coordinates.

Manual plotting of the grid itself may be done with the aid of a T-square and triangle, or a drafting machine. Another good way is to draw two diagonals joining the corners of the sheet and then mark four points on them with a compass set at the intersection (see Fig. 11-1). These points form a perfect rectangle with a marked center within which other grid lines may be added by using a scale and dividers.

Preprinted grid. Sheets of drafting film and vellum are available with a printed grid, often in a no-print or fade-out color. The graticule intersections may be plotted by coordinates using the faint grid lines which disappear when copies of the map are made. Some of the grid lines may be inked if they are wanted.

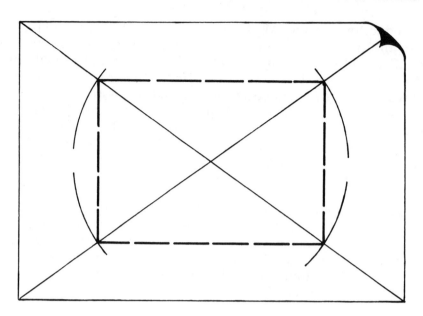

FIG 11-1 Method of centering a rectangle on a map sheet.

Coordinatograph. Special machines, designed for the specific purpose of plotting by rectangular coordinates, may be used if available. See Fig. 11-2a. They
consist of a beam which carries a y scale and plotting head, and rails which guide
its movement along an x scale. The actual scale divisions are on interchangeable
tapes which may be positioned as desired. The plotting head, which moves along
the beam in the y direction, includes a pricking device, pencil chuck, or pen.

Coordinatographs may be used not only to plot the grid lines but also to plot
individual points (survey control points or graticule intersections) and to read
coordinates on existing maps. Polar coordinatographs in which the beam rotates
along a circular rail are available, but they are far less common. See Fig. 11-2b.

Because of their precision and size, coordinatographs are quite expensive.

Flat-bed plotters. The movement of the plotting head of a coordinatograph may
be controlled by a computer program. This is the way that the U.S. Geological
Survey plots the graticules and grid ticks for its quadrangle maps.

Grid templates. The fifth way to lay out a grid is with a metal template. It is
a sheet of dimensionally stable aluminum alloy which has small holes drilled at 4-
or 5-cm intervals. A point-plotting device is centered over each hole to prick the
drawing surface very precisely. The prick marks are then circled lightly and the
grid lines added (see Fig. 11-3).

Templates with 4- and 5-cm spacings are appropriate for a variety of common scales. The former spacing (4 cm) corresponds to 1 km at a scale of 1:25,000;

FIG 11-2a Precision coordinatograph. (Courtesy of ROST-Vienna.)

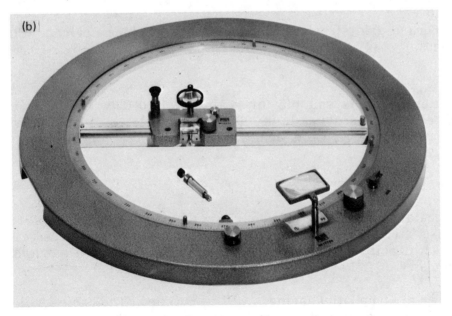

FIG 11-2b Polar coordinatograph. (Courtesy of ROST-Vienna.)

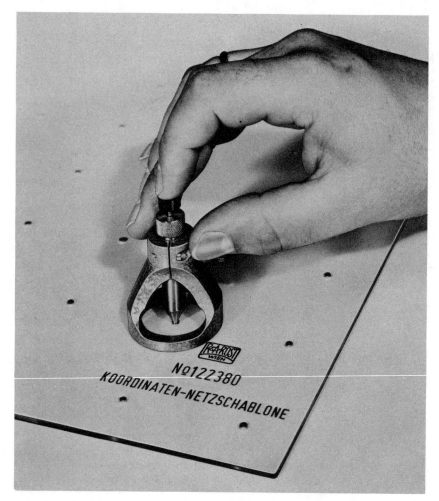

FIG 11-3 Square grid stencil or template with pricking device. (Courtesy of ROST-Vienna.)

and the latter (5 cm) is suitable for scales of 1:10, 000; 1:20, 000; 1:100, 000; 1:200, 000; etc. At a scale of 1:200, 000, for example, the grid interval would be 10 km, and at 1:1, 000, 000 it would be 50 km.

11-3 COMPILATION METHODS

On small-scale maps, as mentioned earlier, the compilation of coastlines, bound-aries, rivers, etc., is derived from existing maps of similar or larger scale and is not based directly on a survey. (The plotting of large-scale original maps by photogrammetry or field surveys will not be covered here.)

A number of compilation methods will be outlined. The choice will depend upon the precision needed, the difference in scale between the source and the product, the equipment available, and whether or not the same map projection was used in both the source and the product.

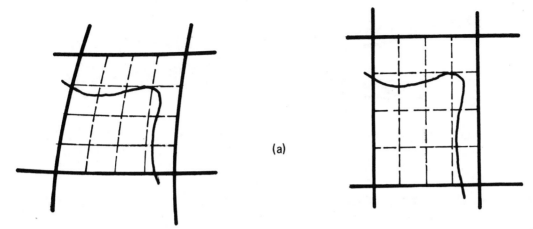

FIG 11-4a Method of compiling "by eye" with the aid of smaller squares temporarily sketched on the source map (left) and the manuscript.

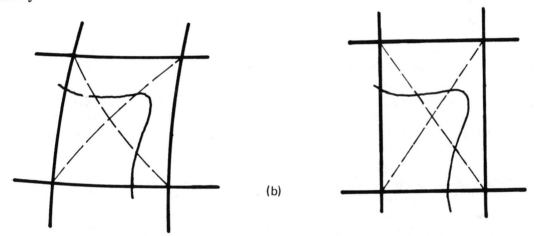

FIG 11-b A variation on the method of Fig. 11-4a in which diagonals are sketched in some of the graticule quads. The source map is on the left.

Photographic reduction. This method does not allow for a selection of what features are to be carried over to the new map and does not permit a change of map projection. Also, of course, the camera does not generalize.

Grid squares. A common method is to break the graticule quadrangle on the existing map into smaller squares and to temporarily do the same thing on the new map. The compiler then judges by eye how a particular coastline should be drawn. He can see which squares are traversed on the original map and attempt to draw a similar configuration in squares of the same interval but of different size and shape on his new map. See Fig. 11-4a.

This method does permit a change of projection, as shown in the figure, and can be quite accurate if sufficient care is taken. A variation is shown in Fig. 11-4b. Instead of dividing the original square into smaller squares, diagonals are added. Again, they serve to aid the eye.

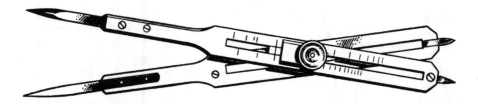

FIG 11-5 Proportional dividers. (From U.S. Army Technical Manual TM 5-240.)

Proportional dividers. This device, shown in Fig. 11-5, is helpful in trans-ferring a distance shown at one scale to a new map of a different scale. It is a double-ended set of dividers in which the pivot point may be adjusted to provide a wide range of ratios between the openings of the two sets of points. When set for a 2 to 1 reduction, for example, the one set of points will always be half as far apart as the other.

Pantograph. This device, not shown, is a mechanical linkage in which a tracer point and a pencil point will trace out the same figure at different scales. It will accomplish a scale change, but not a projection change, of course. The whole ap-paratus pivots about an anchor. The anchor, tracer point, and pencil point lie along a straight line at intervals which are in a fixed ratio as the existing map is traced. The pencil draws the coastline or other feature at the smaller (or larger) scale.

Vertical sketchmaster. The principle of this piece of equipment is shown in Fig. 11-6. It is one of several instruments that enable the compiler to see an image of a source map (or a photograph) superimposed upon his manuscript. It is modestly priced in comparison to the others to be described later. It stands on three legs directly over the part of the manuscript being compiled. The opera-tor looks downward through an eyepiece and sees the manuscript through a semi-transparent mirror (or half-mirror). As he looks at the new map, he also sees a reflected image of the source map and can draw any of its features. The lengths of the legs may be adjusted to allow for a modest difference in scale between the two maps. For a greater difference in scale, various lenses may be used. A slight difference in map projection can be accommodated approximately by tilting the unit and matching images as nearly as possible for a given graticule quadrangle.

Zoom transferscope. This instrument is more elaborate in design than a ver-tical sketchmaster but is similar in purpose. It provides for varying the illumina-tion on the two maps, for rotating the image, for a considerable difference in scale, for stretching the image in one direction, and for using a transparency as a source, but is only capable of enlarging.

Vertical reflecting projector. This is an opaque projector, but instead of pro-jecting an image to a screen as in a classroom, it projects it to a drawing surface.

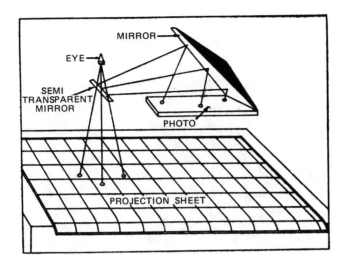

FIG 11-6 Principle of the vertical sketchmaster.   (From U.S. Army Technical Manual TM 5-240.)

Some models, like the one in Fig. 11-7, shine down from above the table, where- as others project the image to the underside of a light table. The former has the disadvantage that the compiler's head and hands will get in the way and cast shad- ows, while the latter requires that he compile on a translucent medium. Both types provide for a considerable difference in scale through adjustments and the use of different lenses.

Digitized data. "Automated cartography" includes the possibility of digitizing an existing map. The computer can then be programmed to redraw it at a differ- ent scale and on a different projection.

11-4 OTHER ASPECTS OF CARTOGRAPHY

This book is limited to the projection of the graticule and geographic features from a globe (a scale model of the earth) to a map. No effort has been made to cover drafting techniques, rules for labeling various classes of features, processing and symbolizing data, map design, or map reproduction. In addition, of course, there has been no effort to cover surveying methods or topographic mapping.

PROBLEMS

11-1 Plot the cylindrical equidistant projection for a map of Africa as described in Example 2-1. This will require a sheet somewhat larger than 20 cm^2.

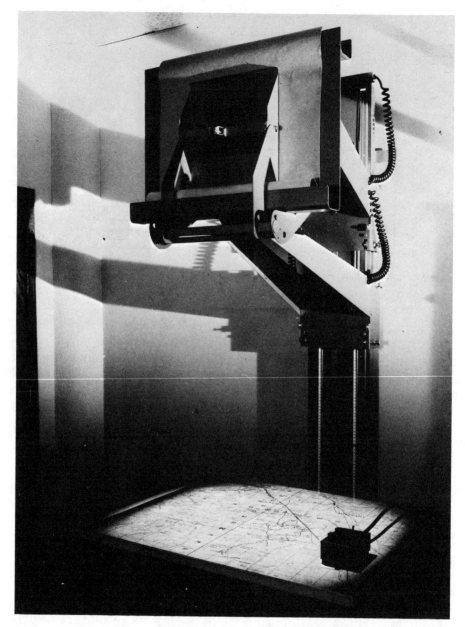

FIG 11-7 Vertical reflecting projector. (Courtesy of Artograph, Inc.)

Lay out the grid with care, using the method involving two diagonals and a
compass (Fig. 11-1). Compile the coastline of Africa using a source map
at least as large as yours, if possible. If you own the source (an atlas or
outline map), you may wish to subdivide the graticule on both sheets. See
Fig. 11-4.

11-2 Prepare a grid, or use a preprinted grid, and plot the graticule computed
      in Example 4-2. This was an Albers projection for Scandinavia. Use the x

and y coordinates shown there.   Compile the coastlines by any suitable method considering what equipment you have available to you and whether the projection used in your atlas was also an Albers or is quite different (having curved meridians, for example).   Ink the drawing if assigned.   Do not label the countries or ocean unless you have studied the conventions used in labeling map features.

Appendix A

PROGRAMS FOR POCKET CALCULATORS

A-1 ALBERS PROJECTION

The following program for a Hewlett-Packard HP-25 pocket calculator will give the rectangular coordinates for plotting an Albers projection. If the meridians to be shown are spaced symmetrically with respect to the central meridian, there will be no need to calculate the western half of the graticule.

Example 4-2 lists the five constants to be stored in advance and shows a table of output from the program. To fill in each column in that table with y and x coordinates, one must store the central angle in STO 0, enter a latitude in the X register, and then start the program. The next latitude is then entered, and so on, until the column is completed. Then a new central angle is stored and the latitudes are entered again in turn. The 18 steps are as follows:

	Key Entry	Comments		Key Entry	Comments
01	f sin	$\sin \phi_1$	10	÷	divide
02	CHS	change sign	11	f $\sqrt{x}$	take square root
03	RCL 4	recall $\sin \phi_n$	12	RCL 0	recall L angle
04	+	add	13	x ⇄ y	exchange x and y
05	RCL 3	recall $2R^2$	14	f R→	to rectangular coord.
06	X	multiply	15	CHS	change sign
07	RCL 2	recall kr_n^2	16	RCL 5	recall r_{max}
08	+	add	17	+	add, display y coord.
09	RCL 1	recall k	18	GTO-00	

(manually exchange x
and y to display x coord.)

The HP-25 could be programmed to do some of the preliminary computations as suggested in Prob. 4-3.

APPENDIX A

A-2 OBLIQUE STEREOGRAPHIC PROJECTION

The following program for an HP-25 pocket calculator will solve the mapping equations given in Sec. 8-3. It displays the x coordinate and holds the y coordinate in the Y register. After the program is entered, $\sin \phi_0$ must be stored in R_0, $\cos \phi_0$ in R_1, ϕ in R_2, and r (= 2R) in R_3. Lastly, λ, the longitude of a meridian with respect to the central meridian, is keyed in and left in the X register. (Actually this is $\Delta\lambda$, a difference in longitude.)

The user should work his way along one parallel at a time changing λ for each run of the program. Then, with a new ϕ entered in R_2, another series of longitudes may be run through the program. South latitudes are negative, of course.

Sample values: If ϕ_0 = +48.5°, ϕ = +30°, r = 45 cm, and λ = 120°, the following coordinates should be obtained:

x = 31.03 cm

y = 27.13 cm

Calculation and plotting work is reduced if the graticule is symmetrical (i.e., the central meridian is chosen so that other meridians will be equally spaced on each side). Appendix A-4 provides a program for this projection (and others) using a Texas Instruments SR-56.

The HP-25 program is as follows:

	Key Entry	Comments		Key Entry	Comments
01	ENTER		22	X	
02	f sin		23	CHS	
03	STO 5	$\sin \lambda$ in R_5	24	RCL 1	
04	R↓		25	RCL 2	
05	f cos	$\cos \lambda$	26	f sin	
06	RCL 2		27	X	
07	f cos		28	+	
08	X		29	RCL 3	
09	STO 7	$\cos \lambda \cos \phi$ in R_7	30	X	
10	RCL 1		31	RCL 6	
11	X		32	÷	
12	RCL 0		33	RCL 5	
13	RCL 2		34	RCL 2	
14	f sin		35	f cos	
15	X		36	X	
16	+		37	RCL 3	
17	1		38	X	
18	+		39	RCL 6	
19	STO 6	denominator in R_6	40	÷	
20	RCL 7		41	GTO-00	x is displayed; y is in Y register
21	RCL 0				

A-3 HAMMER-AITOFF PROJECTION

The following program for an HP-25 pocket calculator solves the equations given in Sec. 9-4. The radius of the generating globe R must be stored in R_0, the latitude ϕ in R_1, and the longitude λ in the X register before the program is started. It is intended that all points on a given parallel will be calculated while ϕ is stored. A numerical example is included in Chap. 9.

The steps are as follows:

	Key Entry	Comments		Key Entry	Comments
01	2		17	X	numerator of K
02	÷		18	x ⇄ y	
03	ENTER		19	÷	
04	f sin		20	STO 5	store K
05	STO 3	store sin $\lambda/2$	21	RCL 1	
06	R↓		22	f sin	
07	f cos	cos $\lambda/2$	23	X	calculate y
08	RCL 1		24	RCL 1	
09	f cos		25	f cos	
10	X	Multiply cos $\lambda/2$ cos ϕ	26	RCL 3	
11	1		27	X	
12	+		28	2	
13	f √x	denominator of K	29	X	
14	2		30	RCL 5	
15	f √x		31	X	calculate x
16	RCL 0		32	GTO-00	x is displayed; y is in Y register

A-4 THE FIVE AZIMUTHAL PROJECTIONS

The following program for a Texas Instruments SR-56 pocket calculator will give the rectangular coordinates for plotting any of the azimuthal projections covered in this book, whether polar, equatorial, or oblique. Steps 13 through 23 must be varied according to which projection is to be calculated. As mentioned in the Preface, the program was contributed by John P. Snyder.

The equations used are

$$x = A \cos \phi \sin \lambda$$

$$y = A (\cos \phi_0 \sin \phi - \sin \phi_0 \cos \phi \cos \lambda)$$

where λ is with respect to the central meridian, ϕ_0 is the central latitude, and

$$A = R/\cos a \text{ for gnomonic}$$

$$A = 2R/(1 + \cos a) \text{ for stereographic}$$

A = R for orthographic

A = Ra/sin a for equidistant (a, called D in Sec. 1-6, is in radians)

A = R $2/(1 + \cos a)^{1/2}$ for Lambert equal-area

$\cos a = \sin \phi_0 \sin \phi + \cos \phi_0 \cos \phi \cos \lambda$, from Eq. (5), Sec. 1-6

Instructions for User

1. Choose the projection to be plotted, the scale in terms of the radius of the generating globe R, the central meridian, and the central latitude ϕ_0. (The central latitude may be at either pole or have any value in between.) Also select the bounding parallels, to be called ϕ_{max} and ϕ_{min}; the bounding meridians, given no special names; and the desired interval between the parallels and meridians to be plotted, called dϕ and dλ. Prepare a sheet of graph paper or vellum having a fade-out or no-print grid for the immediate plotting of the graticule intersections as their coordinates are displayed or, if preferred, prepare a table for listing them for later use.

2. Enter the program of 95 steps, varying steps 13 through 23 as shown according to which projection was selected. Press RST.

3. Enter the variables:

> Enter the radius R. Press STO 0.
>
> Enter the central latitude ϕ_0 in degrees. Press STO 1.
>
> Enter the westernmost longitude as a difference $\Delta\lambda$ from the central meridian (minus if west, plus if east or zero). Press STO 2.
>
> Enter the northernmost latitude to be shown ϕ_{max}, in degrees (this may be the North Pole, 90°). See following notes. Press STO 3.
>
> Enter the southernmost latitude to be shown ϕ_{min}, in degrees (south latitudes being minus). Press STO 4.
>
> Enter the desired interval dϕ and dλ (such as 0.5°, 2°, 5°, or 10°). Press STO 5.

4. Press R/S. After about 8 sec, $\Delta\lambda$ and then ϕ_{max} will blink, then x will appear.

5. Press R/S. Immediately y will appear.

6. Press R/S. The program will automatically increment southward, and $\Delta\lambda$, then ϕ_{max} - dϕ will blink. Then x will appear. Press R/S again and y will appear.

7. Keep repeating step 6. The program will progress southward, blinking $\Delta\lambda$, and the updated latitude, then displaying x and y.

8. After ϕ_{min} is reached, the next press of R/S will automatically increment the longitude to $\Delta\lambda + d\lambda$ and return the program to ϕ_{max} at the top of the next meridian. Those values will blink, followed by x (and y). With successive presses of R/S, the program will work its way southward on the second meridian, then the third, etc., until the user chooses not to continue.

9. For another set of variables on the same projection, press RST and repeat steps 3 to 8.

Notes

On the gnomonic and orthographic projections, which can only show one hemisphere or less, any point beyond one hemisphere or at infinity will be rejected and the program will automatically move to the top of the next meridian and begin calculating visible points. If all remaining meridians are beyond limits, the program will continue searching and the user should press R/S to stop it. If, however, a southern central latitude ϕ_0 is chosen, the ϕ_{max} may not remain within limits on these two projections. In this case, the program will continue to search, but it will not find the more southerly points which are within limits. To prevent this, change steps 17 and 18 from 7, 4 to 6, 1 before starting calculations. This change may be used for all gnomonic and orthographic calculations, but it lengthens the program time near the limits of the map.

The program is not suitable for calculating the point opposite the center ($-\phi_0$, $\Delta\lambda = 180°$), but this is simply a circle on the equidistant and Lambert equal-area and is beyond limits on the other three projections.

Example

Following the eight steps as numbered above, calculate the coordinates for plotting a graticule suitable for a map of the United States.

1. Select the Lambert azimuthal equal-area projection and use $\phi_0 = 40°$, R = 16 cm, central meridian at 96°W. Plot every 10° from 50°N to 30°N and from 120°W to 70°W. The first meridian will be at $\Delta\lambda = -24°$, then $d\lambda$ (10°) will be added until the last meridian is reached at +26°.

2. Enter the program using the special steps under the Lambert column. Press RST.

3. Enter 16 - - - - - - - - - - Press STO 0
    Enter 40 - - - - - - - - - - Press STO 1
    Enter 24, +/- - - - - - - Press STO 2
    Enter 50 - - - - - - - - - - Press STO 3
    Enter 30 - - - - - - - - - - Press STO 4
    Enter 10 - - - - - - - - - - Press STO 5

4. Press R/S. Watch -24 and 50 blink, then -4.2449 will appear. This is x.

5. Press R/S. The y coordinate 3.3994 will appear.

6, 7, 8. On successive presses of R/S the program will proceed as follows:

For Long.	Lat.	Blinking λ	φ	Displayed x	y
120°	40°	-24	40	-5.0497	0.6899
	30	-24	30	-5.7410	-2.0458
110	50	-14	50	-2.5068	2.9972
	40	-14	40	-2.9782	0.2351
	30	-14	30	-3.3818	-2.5360
100	50	- 4	50	-0.7204	2.8060
	40	- 4	40	-0.8553	0.0192
	30	- 4	30	-0.9707	-2.7683
90	50	6	50	1.0799	2.8273
	40	6	40	1.2822	0.0432

... concluding with

70	30	26	30	6.2033	-1.9168

The SR-56 Program:

```
        (LRN)
00    RCL
01     3
02    STO
03     7
04     1
05    x ⇌ t
06    RCL
07     1
08     sin
09    *subr
10     7
11     9
12     cos
```

	Equid.	Lamb.	Ster.	Ortho.	Gnom.
13	=		+	=	
14	*x = t		1	*CP	
15	2		=	INV	+/-
16	4	*1/x		*x ≥ t	
17	*RAD	X		7	
18	INV	2		4	
19	cos	=		1	+/-
20	÷	* √x	*NOP		*1/x
21	sin		*NOP		
22	INV		*NOP		
23	*RAD		*NOP		

```
24     X
25    RCL
26     0
27     X
28    x ⇌ t
29     (
30    RCL
31     1
32     cos
33    *subr
34     7
35     9
36     sin
37    +/-
38     = (y is stored)
39    STO
40     8
41    RCL
42     2
43    INV
44    *fix
45    *pause        λ
46     sin
47     X
48    RCL
49     7
50    *pause        φ
```

51	cos
52	X
53	x ⇄ t
54	=
55	*fix
56	4
57	R/S
58	RCL
59	8
60	R/S
61	RCL
62	5
63	INV
64	SUM
65	7
66	RCL
67	4
68	x ⇄ t
69	RCL
70	7
71	*x ≥ t
72	0
73	4
74	RCL

55 *fix (no. of decimals
56 4 in display.
57 R/S Change to suit.)

75	5
76	SUM
77	2
78	RST
79	X (subroutine starts)
80	RCL
81	7
82	sin
83	+
84	RCL
85	7
86	cos
87	X
88	RCL
89	2
90	cos
91	X
92	RCL
93	1
94	*rtn
	(LRN)
	(RST)

*Denotes "2nd" key.

Appendix B
TABLES

TABLE B-1

Scale Values for Generating Globes (for rough checking of computations)

Radius of Globe in. (cm)	RF Scale (denominator)	Mile Scale, 1 in. =	Graphic Scale, 1,000 miles =	Km Scale, 1 cm =	Graphic Scale, 1,000 km =	Length of 10° Arc in. (cm)
0 (0)	Infinity	Infinity	0.00 in.	Infinity	0.00 in. (0 cm)	0.00 in. (0 cm)
5 (12.7)	50,180,000	792 mi.	1.26 in.	502 km	0.78 in. (1.99 cm)	0.873 in. (2.22)
10 (25.4)	25,100,000	396 mi.	2.52 in.	251 km	1.57 in. (3.99 cm)	1.745 in. (4.43)
15 (38.1)	16,730,000	264 mi.	3.79 in.	167 km	2.35 in. (5.98 cm)	2.618 in. (6.65)
20 (50.8)	12,540,000	198 mi.	5.05 in.	126 km	3.13 in. (7.96 cm)	3.491 in. (8.87)
25 (63.5)	10,040,000	158 mi.	6.31 in.	100 km	3.92 in. (9.96 cm)	4.363 in. (11.08)
30 (76.2)	8,360,000	132 mi.	7.58 in.	83.5 km	4.72 in. (11.98 cm)	5.236 in. (13.30)
35 (88.9)	7,170,000	113.1 mi.	8.84 in.	71.7 km	5.49 in. (13.95 cm)	6.109 in. (15.52)
40 (101.6)	6,270,000	99.0 mi.	10.10 in.	62.7 km	6.28 in. (15.95 cm)	6.981 in. (17.73)
45 (114.3)	5,580,000	88.0 mi.	11.36 in.	55.8 km	7.06 in. (17.92 cm)	7.854 in. (19.95)
50 (127.0)	5,020,000	79.2 mi.	12.63 in.	50.2 km	7.84 in. (19.92 cm)	8.727 in. (22.17)
100 (254.0)	2,510,000	39.6 mi.	25.25 in.	25.1 km	15.70 in. (39.84 cm)	17.453 in. (44.33)

TABLE B-2

Summary of Map Projections

Chapter	Map Projection	Cylindrical, Conic, Azimuthal	Conformal or Equal-Area	Equidistant	Other Characteristics	Applications
2	Cylindrical equidistant	Cyl.	—	N-S	Equatorial case of the following proj. Square pattern, easy to draw.	Small areas near Equator.
3	Cylindrical equidistant with two std. parallels	Cyl.	—	N-S	Rectangles of uniform dimensions, easy to draw. Also called equirectangular projection.	Small areas away from Equator; city or county maps, for example.
4	Cylindrical equal-area	Cyl.	Eq. ar.	—	North-south compression and east-west stretching near poles. Also called rectangular equal-area.	Rarely used. Reasonable for world distributions and areas near Equator.
6	Mercator	Cyl.	Conf.	—	Loxodromes are straight; pole is at infinity. Scale factor equal to sec φ.	Navigation. (Also see transverse case below, useful in surveying.)
7	Gnomonic cylindrical	Cyl.	—	—	Rectangular graticule; cannot show poles. Has no great merit.	No importance.
7	Perspective conic (gnomonic)	Conic	—	—	Has no great merit.	No importance.
2	Conical equidistant	Conic	—	N-S	Parallels are concentric circles, pole is an arc. Called simple conic.	Other conics are better.
3	Two-standard conic	Conic	—	N-S	Better scale than simple conic.	Other conics are better.
None	One-standard equal-area conic	Conic	Eq. ar.	—	East-west stretching and north-south compression.	Other equal-area conics are better.

TABLE B-2 (continued)

Chapter	Map Projection	Cylindrical, Conic, Azimuthal	Conformal or Equal-Area	Equidistant	Other Characteristics	Applications
5	Bonne	Conic*	Eq. ar.	E-W	Curved meridians. Similar in theory to sinusoidal.	Square areas such as France and Switzerland.
4	Albers	Conic	Eq. ar.	—	Straight meridians. Two standard parallels.	Ideal for continental U.S. (wide in the east-west direction).
5	Polyconic	Conic*	—	E-W	"Many cones," each with its own std. parallel. Several variations are used.	Large-scale maps in series (topographic) with several central meridians.
6	Lambert conformal conic	Conic	Conf.	—	Concentric parallels and radial meridians. Two standard parallels.	Aeronautical charts, meteorology, state plane coordinates in many states.
6, 8 (10)	Stereographic	Az.	Conf.	—	Circles on globe remain circular. Oblique case may be accurately constructed with beam compass.	Oblique transformations, to form other azimuthals. Military grid near poles (UPS). Hemispheres.
5	Orthographic	Az.	—	E-W on polar	Looks like a photo of a globe; shows a hemisphere or less.	Illustrations, art work.
7	Gnomonic	Az.	—	—	Great circles are straight lines. Shows less than a hemisphere, easy to construct.	For navigation, in conjunction with Mercator.
4, 8	Azimuthal equal-area	Az.	Eq. ar.	—	Radial compression, tangential stretching.	Hemispheres or continents.

2, 8	Azimuthal equi-distant	Az.	—	Radially	Distances and directions from center are shown correctly.	Air travel from center, earthquake studies, radio transmission.
10	Globular	—	—	—	Better average scale in a hemisphere than orthographic and stereographic. Easy to draw.	Hemisphere maps, especially in schools.
10	Transverse Mercator	Cyl.	Conf.	—	Looks peculiar for whole world; used for areas with large dimension north and south.	Military grid (UTM). State plane coordinates in many states.
5	Sinusoidal	Cyl. *	Eq. ar.	E-W	Meridians are sine curves.	World, Africa, S. America.
9	Mollweide	Cyl. *	Eq. ar.	—	Elliptical meridians, rounded poles.	World distributions.
9	Hammer-Aitoff	Az. *	Eq. ar.	—	Curved parallels, better angles between meridians and parallels.	World distributions.
None	Eckert IV	Cyl. *	Eq. ar.	—	Pole is a line half as long as Equator, similar to flat polar quartic.	World distributions.
None	Flat polar quartic	Cyl. *	Eq. ar.	—	Pole is a line one-third as long as Equator. Better appearance in polar area than sinusoidal, etc.	World distributions.
9	Goode's Homolosine	Cyl. *	Eq. ar.	—	Combination of sinusoidal and Mollweide along 40th parallel.	Interrupted world maps for distributions.

Note: The term equidistant is not commonly applied to projections with all parallels standard, as is done here.

*Pseudo

Appendix C

TISSOT'S INDICATRIX

Chapter 1 includes a brief discussion of how a tiny circle on the generating globe is deformed during the process of projection. In all cases, the circle projects as an ellipse. If the projection is conformal, the ellipse is circular, having the same shape as before projection but generally not the same size. On an equal-area projection, because of compensatory scale factors, the ellipse will be noncircular in most parts of the map and will be smaller than the circle in one dimension and larger in the other. In the case of all other projections, the ellipse will in most locations be noncircular and also of larger or smaller area than the original circle.

The major and minor axes of the projected ellipse are perpendicular, of course, and also were perpendicular prior to projection. The 90° angle between them was not deformed but, except for the case of a circular ellipse, other angles are deformed.

The discussion presented in Chap. 1, and reviewed in some of the numerical problems in other chapters, was based on Tissot's indicatrix, an indicator of map deformation developed by M. A. Tissot in France in 1881.

C-1 THEORY OF THE INDICATRIX

The radius of the small circle on the globe is assumed to be 1.000 and represents the scale factor (the circle is theoretical, not finite). In the ellipse, the semimajor axis represents the maximum scale factor at the particular point being considered and the semiminor axis represents the minimum scale factor at the same point. The maximum is called a and the minimum is called b. The product ab is called S, and serves to indicate how much area exaggeration exists at that location. If S = 1.000, there is no exaggeration. If it is 1.000 at all points, the projection is equal-area. If it is 2.000 at some point, a local feature will appear twice its proper size.

158

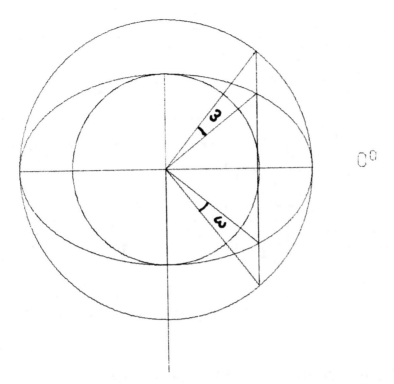

FIG C-1 Tissot's indicatrix for a point on the Equator on a polar azimuthal equidistant projection. Plotted by computer at Indiana State University, Terre Haute, Indiana. (Courtesy of Drs. William D. Brooks and Charles E. Roberts.)

On a conformal projection, a = b at all points on the map. The product S will vary, of course. On a Mercator projection, a and b both equal 2.000 at the 60th parallel and the area exaggeration there is 4.000.

An angular deformation exists at a point if there is any difference between a and b. The maximum change in the direction of one radius of the circle is called omega (ω) and the maximum change in an angle between two radii is therefore 2ω. If a and b are known, 2ω may be calculated as follows:

$$2\omega = 2 \arcsin \frac{a - b}{a + b} \tag{1}$$

A derivation of this expression is given in Ref. 1.

Figure C-1 shows a plotting of Tissot's indicatrix for a point on the Equator in a polar azimuthal equidistant projection. Radially, the scale factor is 1.000 in this projection. Along the Equator the scale factor is the "map distance" $2\pi(\pi R/2) = R\pi^2$ divided by the "globe distance" $2\pi R$.

$$\text{Scale factor} = \frac{R\pi^2}{2\pi R} = \frac{\pi}{2} = 1.571$$

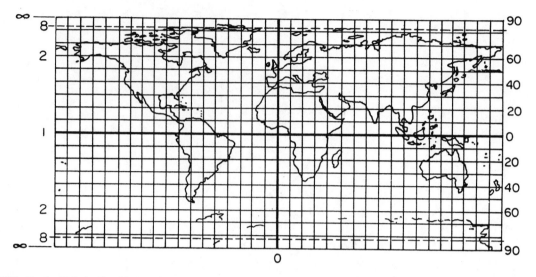

FIG C-2 Distribution of area exaggeration on a cylindrical equidistant projection. Lines are shown for S = 1, 2, 8, and infinity. Angular deformation also exists.

Thus $\underline{a}$ = 1. 571, $\underline{b}$ = 1. 000, and S and 2ω are as follows:

S = $\underline{ab}$ = (1. 571)(1. 000) = 1. 571

$2\omega = 2 \arcsin \dfrac{1.\,571 - 1.\,000}{1.571 + 1.\,000} = 2 \arcsin 0.\,222 = 25.\,66°$

The figure shows the unit circle in the center surrounded by the oval-shaped, or elliptical, indicatrix, plus an enveloping circle with a radius of 1.571. The two radial directions which suffer the most angular deformation are shown extending to the large circle as they would do if the projection were conformal ($\underline{a} = \underline{b} =$ 1.571), and in their deformed positions, extending to the indicatrix. The two angular deformations ω are shown. Each is equal to 12. 83°.

C-2 OTHER ILLUSTRATIONS

Figure 1-15 shows the distribution of angular deformation on a sinusoidal projection. Lines showing where $2\omega = 10°$ and 40° are shown.

Figure C-2 shows the distribution of area exaggeration on a cylindrical equidistant projection. Lines showing where S = 1, 2, 8, and infinity are shown.

Example C-1

Calculate where the maximum angular deformation reaches 40° on a cylindrical equidistant projection.

Solution

On this projection, $\underline{b}$ = 1.000 everywhere, and $\underline{a}$ varies with sec ϕ. The first step is to find $\underline{a}$ for the location having the 40° deformation.

$$2\omega = 40° = 2\ \text{arcsin}\ \frac{a - 1.0}{a + 1.0}$$

$$\sin 20° = 0.342 = \frac{a - 1.0}{a + 1.0}$$

$$0.342a + 0.342 - a = -1.0$$

$$-0.658a = -1.342$$

$$a = 2.040 = \sec \phi$$

$$\phi = 60.6°$$

Lines of equal deformation are circular on all azimuthal projections and are parallel to the standard parallels on cylindrical and conic projections. Their pattern is more complex on pseudocylindrical, pseudoconic, pseudoazimuthal, and other projections not closely related to the developable surfaces. This was shown in Fig. 1-15 (sinusoidal). Other examples are shown in the book from which that figure was taken [1] and in papers on map projections in The American Cartographer and other journals [2]. In some cases the computation of $\underline{a}$ and $\underline{b}$ is more difficult than it was for Fig. C-1 or Ex. C-1, involving partial derivatives.

Two more examples of Tissot's indicatrix are shown in Figs. C-3 and C-4. Like the one in Fig. C-1, these were plotted by a computer.

REFERENCES

1. A. H. Robinson and R. D. Sale, Elements of Cartography, John Wiley and Sons, 1969.

2. W. D. Brooks and C. E. Roberts, Jr., The American Cartographer, Vol. 3, No. 2, 1976.

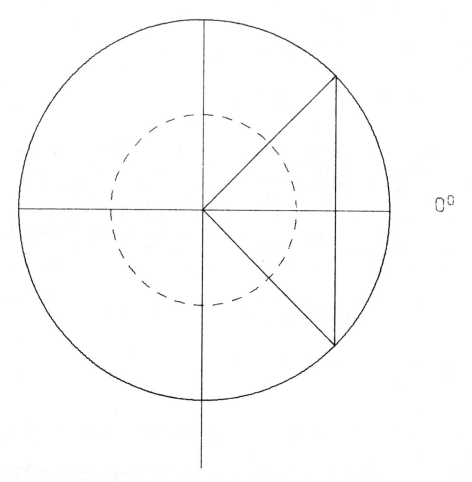

FIG C-3 Tissot's indicatrix for a point on the Equator on a polar stereographic projection. Because the projection is conformal, the indicatrix is a circle (the larger one shown). Scale factors a and b are 2. 00 and S = 4. 00. (Courtesy of Drs. William D. Brooks and Charles E. Roberts.)

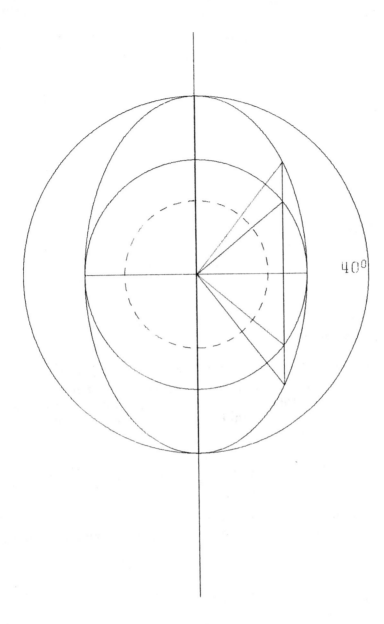

FIG C-4 Tissot's indicatrix for a point on the 40th parallel on a polar gnomonic projection. The dashed circle represents a scale factor of 1. 00.   Scale factors are a = 2. 420 in the north-south direction and b = 1. 556 along the parallel. Maximum angular deformation occurs between the two directions represented by the shorter radial lines. They are each deformed by -12. 56° to the directions of the radial lines which extend to the ellipse. Thus 2ω = -25. 12° and S = 3. 766. (Courtesy of Drs. William D. Brooks and Charles E. Roberts.)

Appendix D

SPHERICAL TRIANGLE
PLOTTED ON VARIOUS PROJECTIONS

Figures D-1 to D-10 show how a certain spherical triangle appears when plotted on 10 different projections.

Great circles joining Anchorage, Alaska; Madrid, Spain; and Buenos Aires, Argentina were plotted on a large globe using a grease pencil. Students then transferred the triangle to their graticules by reading a series of ϕ and λ values along each great circle. The original drawings were made at a much larger scale.

The central meridian for each graticule is 90°W, and the meridians and parallels are spaced at either 15° or 30°. (Intermediate ones were used but erased.)

The drawings shown here were made by Richard Byrem, Willard Yarnall, Chris Lamison, David Beideman, Robert Kline, and Douglas Brehm.

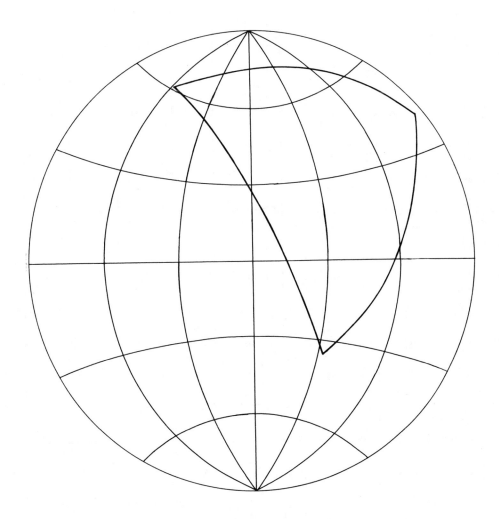

FIG D-1 Globular projection.

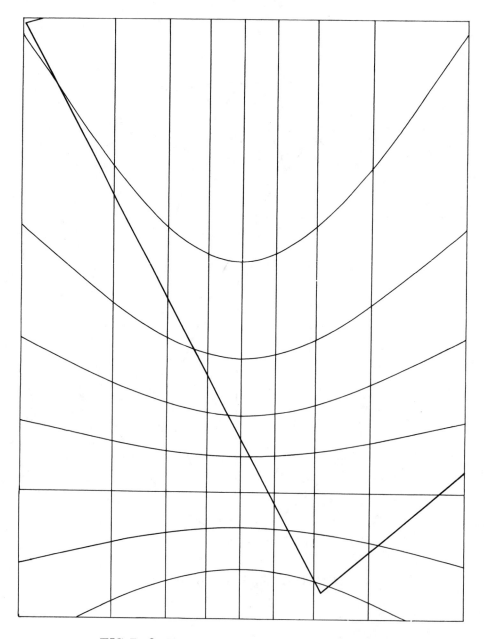

FIG D-2 Equatorial gnomonic projection.

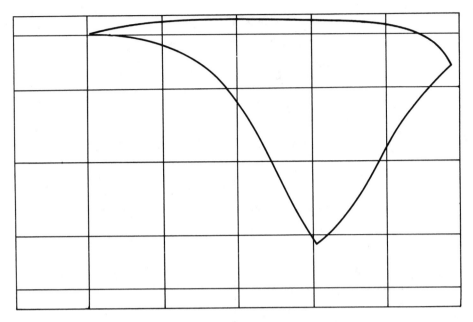

FIG D-3 Cylindrical equal-area projection.

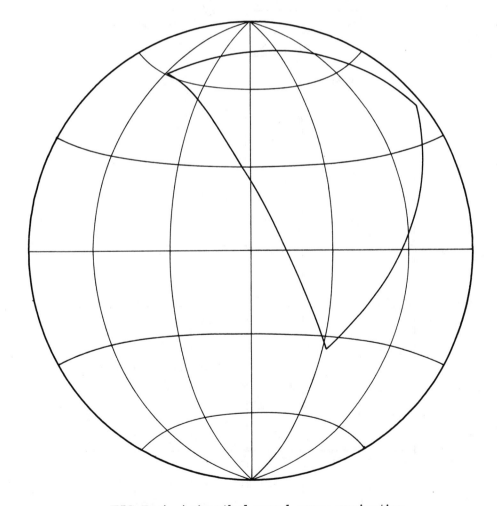

FIG D-4 Azimuthal equal-area projection.

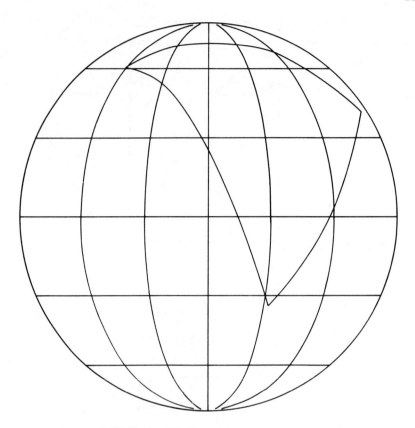

FIG D-5 Mollweide projection.

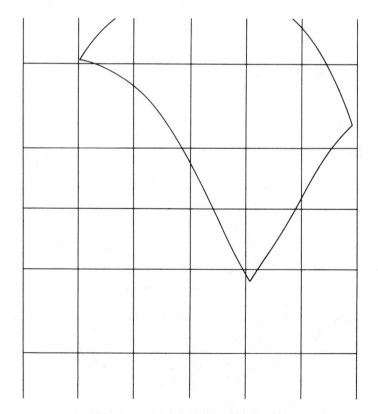

FIG D-6 Mercator projection.

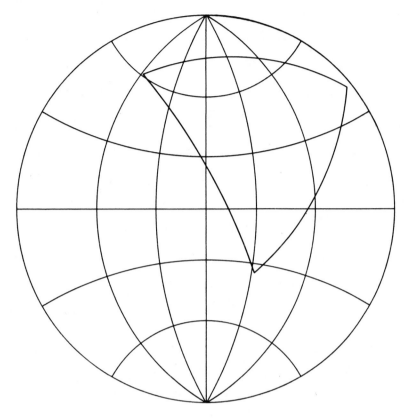

FIG D-7 Stereographic projection.

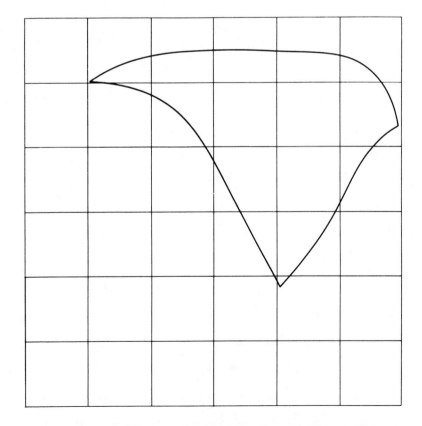

FIG D-8 Cylindrical equidistant projection.

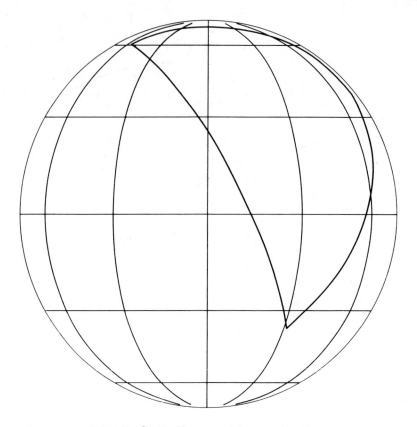

FIG D-9 Orthographic projection.

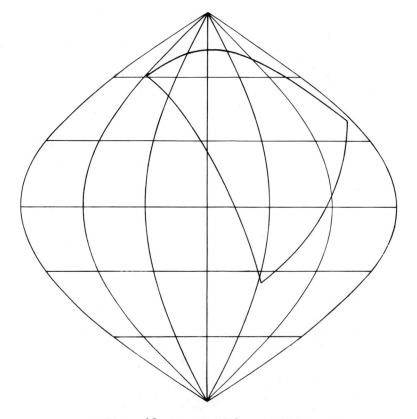

FIG D-10 Sinusoidal projection.